Beautiful Rocks & How to Find Them

A Modern Rockhound's Guide

Beautiful Rocks & How to Find Them

A Modern Rockhound's Guide

ALISON JEAN COLE

PA PRESS

PRINCETON ARCHITECTURAL PRESS · NEW YORK

Published by
Princeton Architectural Press
A division of Chronicle Books LLC
70 West 36th Street
New York, NY 10018 www.papress.com

© 2024 Alison Jean Cole
All rights reserved.
Printed and bound in China
27 26 25 24 4 3 2 1 First edition
ISBN 978-1-7972-2443-5

Editor: Holly La Due
Designer: Natalie Snodgrass

Library of Congress Control Number: 2023942968

Easy.
This one's for Deb.

Introduction

It was a hot day that turned into an almost infernal day. That happens easily out in the badlands, where few plants grow and the earth acts like an oven. My wife, Lisa, our trusty dog, Lilo, and I were out exploring the eroding hills of ancient lake sediments in Utah's Uintah Basin. These sediments, laid down by a massive lake about 50 million years ago, now sit dry as a bone on the earth's surface. The result is an otherworldly landscape of colorfully banded hills, known as badlands, as far as the eye can see. Underneath them lies a massive reserve of oil and gas, the remnants of everything that lived and died in this long-gone lake. The Uintah Basin is a place that should be a national park, but instead it's an ocean of oil derricks and pump stations spanning hundreds of square miles.

A few days prior, I had been perusing satellite imagery of the basin, curious about its badlands and their remoteness. Because the endless tangle of oil roads made the area seem magically accessible by car, I convinced Lisa that we should go explore it. Now, here we were, navigating the labyrinthian maze of dirt roads, passing gas vents that burped plumes of fire into the roasting air. Lilo panted heavily in the car as I gazed out upon the industrial hellscape, wondering why Lisa trusted me to take us anywhere. I reminded myself that the gas in our tank came from this place, or another just like it, and turned my attention back to the map I had made of the area. In my research I had found a cluster of badlands that harbored a secluded cove, and it seemed like a great place to camp for the night. At least we might be able to tuck ourselves out of view from the oil fields. To throw a sheet over the proverbial mirror, so to speak, and pretend to live in the nonhuman world for a night.

Rock hunting, or rockhounding, is always the ulterior motive on any trip we take. We were uncertain what we'd find out here, if anything. However, we'd learned from trips to other badlands environments that they often yield some sort of treasure. The day was getting long in the tooth as we zig-zagged a confusion of dirt tracks to find the site. Finally, we passed the last pump jack and the road ended, tossing us towards freedom. We drove our vehicle across a sandy ditch and overland toward the brightly banded hillsides in search of our dry, dusty cove. We parked at the mouth of it and jumped out, only to be greeted by a sweltering mass of desert air and swarms of biting flies. Not a breeze in sight. I looked at the hills of ancient lake sediments. Had we arrived 50 million years ago, we could have at least gone swimming. To make mat-ters worse, I had insisted we leave the tent at home so we could sleep in the new car. I had just bought a used Toyota 4Runner and I wanted to see if we could all sleep comfortably in the back. (Lilo is 93 pounds, and the answer, we discovered, was no.) This meant windows down for air and bugs inside all night. There's always something humbling to learn on these trips that makes you a better traveler in the future.

Lisa, ever positive, insisted we take a brisk walk into the hills despite the swarms and try to enjoy ourselves before sundown. We scurried along the edges of the eroding hillsides, swatting at bugs, examining material on the ground. At first it was mostly bullet casings and some plastic water bottles. We clearly weren't the first to kill some time here. Within a few paces, however, things began to improve. Lisa started picking up strange fragments of a deep purple stone, saying, "This looks like bone." I was skeptical at first until I heard her gasp a few paces farther on. I looked up to see her holding out a distinctive limb bone. Heavy, dense, and fossilized. Beautiful, undeniable, and old. We followed the fragments up slope, finding more pieces as we went. At this point in our rockhounding hobby we knew that fossil bones were protected by law, so we picked them up, took pictures, and placed them back on the ground. I found a small fragment that was undoubtedly part of a skull and almost

wept with delight. I heard Lisa shout to me, which I always love, because it means she's found something *really* good. There she stood, at the edge of the hillside, pointing to what looked like the profile of a complete skull protruding out of the sediment. It was about the size of a football, and there seemed to be more bones embedded in the layer alongside it.

What are the chances that someone would see interesting hills on satellite imagery, single out one of those hills that looked fun to camp near, drive all day to get to it, park below it, walk up to it, and find such a bounty of fossils? The chances are not as low as you'd think. The evening we found these fossils was truly a lucky one, but it was years of developing a sense for where to find interesting rocks that led us there. We knew from experience that these places yielded interesting material—not just fossils but beautiful rocks and minerals too. About five years prior to this, I wouldn't have had the gumption to navigate the maze of dirt roads in the Uintah Basin, let alone be able to identify a dark purple rock as the fragment of a large vertebrate animal from deep in the past. I would have walked right past that rock. That fossil. That magnificent beast that once sauntered along the edge of a massive lake. I wouldn't even have known that such a lake ever existed, and that one day it just up and vanished. I wouldn't have been able to appreciate the story of the earth in the way that I can now.

For many, rockhounding is a hobby purely concerned with the aesthetic beauty of a rock. People seek material out solely based on its color, its shine, its patterns, its value. Many rockhounds couldn't tell you the composition of the material they collect or much about its journey through time. It makes me sad to know just how much amazing information they're missing out on. Especially since this kind of information is readily available if you're willing to look for it. I wrote this book because I believe that a rock's story is its most valuable quality and that a willingness to research that story makes us more mindful and more appreciative of what we collect. I want to help us all become better rockhounds. I want to help you see the earth! Not just as it is today, but as it was in the past, and the past before that, and the past before that one too.

A good rockhound is a good researcher. Learning to understand the geology of a place will lead you to the best rocks. However, there are cultural overlays that sit atop the geology of our world and it's essential for rockhounds to work harmoniously within them. These cultural overlays are our political maps—public and private lands. All rockhounds need to know what kind of land they're on and what the collecting rules are ahead of time. There are ramifications for trespassing on private property or collecting prohibited items on public land that can really ruin your day.

Learning to read the landscape both geologically and sociopolitically also helps make us more sensitive to the truth that the United States and Canada are merely a veneer upon historic Indigenous lands, most of them taken by force and through broken treaties. North America has a long history of taking. And rockhounding is rock-taking. I consider this heavily in my own collection practice. If I take this rock, what is its future? What will I make with it? Will I display it or gift it in a meaningful way? Is the rock better off with me rather than here in this place? If the rock is destined for a bucket in the dark corner of the garage, then the answer is no. One of the ways my rockhounding hobby has evolved over the years is that sometimes I don't take any rocks home. But I still have just as much fun finding them. Collecting with intention is the single most important practice a rockhound can employ.

Our time in the Uintah Basin ended up being a rockhounding trip where we couldn't take anything home even if we wanted to. It was a blessing that we were well versed in fossil collection rules when we came upon this site. When we first started out in the hobby, we were unaware of collecting rules and were poor identifiers of material. It's likely we would have taken some of these strange purple rocks home where they would have lived on a shelf while we remained oblivious to what they really were. A sad fate for both the rock and the people. Luckily, we were older and wiser rockhounds on this sweltering evening in the badlands. Lisa and I took dozens of photographs of the bones embedded in the hillside and noted our coordinates on our GPS. Neither of us are paleontologists, so we hadn't a clue what kind of large animals we had happened upon. All we knew was that the surface of the earth had worn many different faces between our lifetime and theirs.

Had this location been truly remote, as in a place humans would be unable to drive right up to, we probably would have left it alone. However, the human trash and bullet casings lying just yards from the fossils made it seem likely that eventually someone would probably come try to dig them out. We decided to see if we could help make sure the right people did it. I started getting in touch with paleontologists who worked on digs in Utah. I shared my coordinates and images with them. At first no one was interested. Then, somehow, about a year later, paleontologists and researchers from the local field house in Vernal, Utah, went to dig them up. They informed us that Lisa had found two large turtles, one almost complete, as well as fragments from a brontothere. Imagine a hulking two-horned rhinoceros plodding along the edge of a reedy lakeshore. That's our guy, the brontothere! We were over the moon. It felt like our dedication to the hobby of rockhounding was really starting to pay off, even though we didn't take any rocks home.

Sometimes the reward of rockhounding evolves beyond the initial goal of collecting material. However, the truth is that when we set out on a mission to find interesting rocks and minerals, we're hoping to come home with something special. This book is written with that mission directly in mind. The pages ahead lay out how to find beautiful and interesting material anywhere you live or travel. There's a new way to think about rockhounding—it's all about developing a deeper understanding of *place* and a greater appreciation of the impact rock-taking has on it. When I say place, I mean *the earth beneath your feet*. What is it made of? What's its story through time? Who controls it now and what are you allowed to do on it? The purpose of this book is to help you answer those questions. Perhaps you've been collecting rocks since you were a kid. Or maybe you are completely new to the hobby! Either way, the information presented here is worthy of your time.

I've been working as a professional lapidary artist (cutting stones for jewelry) for the last decade and work mostly with rocks I've collected myself. Often when I go rockhounding, I am looking for material to use in my craft. Since not everyone who reads this book will be a jeweler or lapidary artist looking for craft-quality material, I want you to hear from people who go rockhounding purely for the joy of it. In these pages, you'll find rockhounds from all over North America sharing their strategies, misadventures, and rock collecting philosophies.

You'll hear from the rockhounding community about how to uncover the geological history of the places you intend to hunt, where collecting material is allowed, and creative ways to appreciate your find once you get it back home. We'll get into the nuances of learning to identify material in the field and how to avoid collecting protected things like vertebrate fossils and artifacts. Whether you're into beach pebbles from Cape Cod, emeralds from the Appalachians, gold nuggets from Yukon rivers, snail fossils from Wyoming's agate beds, or geodes dug from Iowa mud, this book will help you become an intrepid and resourceful collector. It will guide you through a new way to navigate the earth's surface by taking a headfirst dive into the deep well of time.

CHAPTER 1

Beautiful Rocks Are Everywhere

"Rocks are reminders that history,
wonder, beauty, and surprise are everywhere,
all around us, all the time."

—Nora Bauman, rockhound in Oregon

No matter where you live in North America, beautiful and interesting rocks are certain to be found nearby. Just follow paths of erosion and you'll find yourself a bounty. Wind, water, gravity, and time all conspire to topple the earth's mountains, pitching the rubble into valleys and dragging it all out to sea. Even if you live amid the Great Plains, where the battle to flatten the earth has largely been won, there are still whispers of erosion where rock material sits exposed at the surface. Seasoned rockhounds from all parts of North America know that they need to make their way to places where there is enough erosional tumult to prevent soil and plant life from carpeting the rocks below. These places take the form of beaches, rivers and creeks, mountain slopes and hillsides, canyons, arroyos, and desert valleys. In some places mankind has lent the earth's forces a hand, blasting quarries and tunneling mine shafts, leaving a path of exposed rock in their wake.

A rockhound's most important skill is building a strong intuition about the kinds of places where rocks show up. While rockhounds need to be savvy about where they're allowed to collect, developing a good sense for where to find material in the most basic sense is the foundation of the hobby. If you can identify a place near you that is likely to have a lot of rocks, you can then research its accessibility. Of the varied environments where rocks show up, some yield a greater variety of material than others. All are worthy of your exploration. The best way to understand these kinds of places is to think about how rocks end up there to begin with. In chapter 2 we'll do a deep dive into what *kinds* of rocks show up in different parts of North America, but for now let's take an eagle-eyed view of the best places to collect material in general.

Natural Collecting Environments

MOUNTAIN SLOPES & HILLSIDES

✦✦✦✦✦

Mountain slopes and hillsides make excellent collecting locations when rock is exposed. They can be strenuous to traverse, but worth the reward. These places are often the *source* of the rock material we seek. Source is an important notion in rockhounding. It's where material breaks off from the motherload and begins its journey downhill. The rocks weren't necessarily born in this place, but it's where they ended up as the Earth shifted its coat around.

The Earth's crust is broken into fragments, or plates, and they split, grow, sink, and collide in a tumultuous dance. In places where they collide or slip underneath each other, large swaths of the crust can be physically forced up, or volcanically burped up, into high ranges. Gravity, water, wind, and time race to knock them down in perhaps the longest and strangest game of Whac-A-Mole ever played. It's not a linear process though. "Mountains are not somehow created whole and subsequently worn away," writes John McPhee in his geology epic *Basin and Range*. "They wear down as they come up."

Many rockhounds consider mountain slopes and hillsides to be excellent

collecting locations because of the abundance of material in close proximity to the source. Prospecting for precious metals and gemstones often happens on slopes since mountain-building events can push large deposits from deep in the crust up to the surface—a migration that takes millions of years. Once humans discover a deposit, it can be mined out in the blink of an eye.

VALLEYS

✢✢✢✢✢

Valleys are the first stop on a rock's journey downhill. In many cases they are also the final stop if a river doesn't sweep up material and send it further along. In environments with a lot of rainfall, valley rocks are often buried under soft sediments, which give rise to a lush carpet of life. Think forests, meadows, and farmlands. Good rock material can be hard to find in these verdant places. In drier climates, however, a lot of rock material remains exposed at the surface. There's not enough rainfall to support the building of deep soils. In the southern deserts and northern tundras, valleys can be good places to collect, but material will be dispersed and not nearly as abundant as it is at the source. When rockhounds are out scanning valley floors for interesting rocks, they refer to the material scattered on the ground as "float." It means the rock or fossil is floating on the surface and you don't have to dig for it.

RIVERS & CREEKS

✦ ✦ ✦ ✦ ✦

Rivers and creeks are the next favorite for rockhounds across the continent. No matter what region you live in there will be waterways nearby. Rivers, creeks, brooks, and streams are conveyor belts of rubble getting dragged off the continent and out to sea. They hold the greatest variety of material of all rock collecting environments. Think of a river as a library of every geological layer the waterway has passed through—"a ransacked library" as my friend and geologist Martin Holden would say. Amid the chaos of the cobbles and gravels are the geological histories of your region.

Not all rivers and creeks are fruitful collection places. Experienced rockhounds have learned to read the rivers they live near. They know that steeper and faster flowing rivers have larger rocks. They know that rivers running through flat territory yield finer gravels and sands. They know that the water level can change dramatically after snow melts and rainfalls. That summer stillness allows algae to carpet

1776—Eagle Cave, Peninsula State Park, Door County, Wisconsin

the river rocks, making them harder to see (and slippery to walk on!). And almost all river rockhounds will tell you that the best material is "always on the other side!" These waterways are dynamic places, changing with the seasons, with new material slowly tumbling its way downstream.

DESERT CANYONS, WASHES & ARROYOS

✢✢✢✢✢

Canyons, washes, and arroyos are the waterways of the deserts. Canyons are incised into the earth with towering walls overhead. Draws, washes, and arroyos tend to be broader and shallower, sometimes braiding themselves through the open desert. Most of them are dry year-round—until they aren't. These quiet, dusty riverbeds are conduits for the unfathomable volume of water that drains into them after thunderstorms and summer monsoons. One moment they are a serene place frozen in time, the next they become what my auto insurance company once deemed "an act of God." (In college my beloved '79 Volkswagen Rabbit was carried off in a flash flood. They sent me a letter blaming the heavens and a check for $500.) These catastrophic events are the calling card of desert rivers. In his exhilarating book *The Secret Knowledge of Water*, explorer Craig Childs reminisces on the mercilessness of these seemingly rare events. "Now come the floods. They charge down atavistic canyons drinking furiously out of thunderstorms, coming one after the next with vomited boulders and trees pounding from one side of a canyon to the other, sometimes no more than hours apart. Sometimes a hundred years apart. Sometimes a thousand. The floods always come."

Perhaps these descriptions are enough to scare you away, but the truth is that these fleeting events are what make desert waterways such fantastic collecting locations. Flooding events sweep massive loads of fresh rock material downstream, totally transforming the riverbed. Of all the possible places to search for rocks, desert washes are my favorite. I love the variety of material I can find in them. I love the slow meandering trek upstream. For the great majority of their epic lives, canyons, washes, and arroyos are quiet places brimming with beautiful rocks. A rockhound's responsibility is to know the incoming weather and always watch the sky. If a cloud even hints at rain, it's time to go.

BADLANDS

✢ ✢ ✢ ✢ ✢

Badlands are a distinct type of desert environment where the surface of the earth is composed of ancient sediments that managed to escape the fate of becoming a solid rock. Think of them as a soft and old geological layer. Something to be tender with. Badland formations are the remnants of ancient valleys and lake basins. Today they are just mounds of silt and clay, slowly washing away with each rain. Badlands often yield rock material, especially in the form of river rocks from extinct streams. Plant and animal fossils are often found in the ancient muds. The best practice is to search among the washes and flats below the hills, and not to trod up them, disturbing their delicate surfaces.

BEACHES & COVES

✢ ✢ ✢ ✢ ✢

In the battle to flatten the earth, the rock that makes up the continents is ground down by the elements and carried out to sea. Our beaches are the final resting place

for all the debris ferried down by rivers and streams. Rocks and sand settle out on the shores, tossed back and forth in the infinite rhythm of waves. Only fine clay particles make it farther on the journey; they drift in water like leaves on a breeze. These fine particles are swept far out on ocean currents and settle slowly over the millennia to the abyssal plains at the bottom of the sea.

Stony beaches and rocky coves are beloved collecting places for coastal rockhounds. For some, even sandy beaches provide collecting opportunities. There are grains of sand out there as old as the planet itself. "The world's beaches faithfully record a detailed history of our planet's geological past," says sedimentologist Milo Barham, "with billions of years of Earth's history imprinted in the geology of each grain of sand." Barham and his team search Earth's beaches for zircon crystals captured within grains of sand. They use the zircon as a deep-time scrying mirror. A geologist's crystal ball.

Most rockhounds, though, are looking for something bigger than grains of sand. This is where good research skills come in handy; we're on the hunt for rocky beaches and coves. Thanks to wave action, the stones on rocky beaches are beautifully polished over the eons by abrasive jostling among other rocks and sand. There is often an extraordinary variety of material on rocky beaches thanks to the myriad of rivers that dump their ransacked library of earth into the sea. I once found a cove on Lopez Island in the San Juan Islands of Washington that was a literal rainbow of rocks. It took my breath away! Even the shores of the interior Great Lakes have rocky stretches that yield some of the most sought-after rocks in North America—banded agate and also sodalite that fluoresces brilliantly under UV light. Rockhounds scan the beaches on moonless nights in search of the glowing material.

Man-Made Collecting Environments

ROADCUTS

✣ ✣ ✣ ✣ ✣

Roadcuts receive a lot of attention in rockhounding, especially for folks with a particular interest in geology. These places are large cuts in the earth where humans have blasted a path through a slope to make room for a roadway (see opposite page). You've likely driven through many roadcuts in your life, especially along highways. After blasting, a beautiful cross section of the bedrock

is exposed. Sometimes veins of metal ores and crystal pockets are revealed in the bedrock. Collecting rules for roadcuts, which are legally considered right-of-ways, vary tremendously among county, state, and federal road maintainers. Sometimes, explicit rules don't exist for roadsides at all. When they do exist, they can be difficult to track down.

I consider roadcut collecting to be rather dangerous, and I want to advise you against it. Roadcuts are cleared to make way for a more level passage through an area. Room is rarely made for safe parking. Recreation along the exposed cliff is never considered when the site is engineered. In some cases, the exposed rock walls are reinforced, but in most cases, especially with older roadcuts, they are left totally precarious. Fractures from blasting permeate the rock, rendering it unstable. Take a hammer to it and you take a big risk.

MINES

✧✧✧✧✧

Mining is humanity's most primal industry. From the earth we pull everything we need. Not only do we mine the bedrock for metals and minerals, but we also mine the soil for its nutrients and the forests for its trees. We drag helium from the air and salt from the oceans. We fill our teeth with gold. We fill our museums with the dead. No stretch of earth is safe from our shovels.

In the vast world of mines, there are two categories that are friendly to our hobby: active mines that serve rockhounds specifically and abandoned mines from a bygone era. The first type is usually much safer than the other. All over North America you will find dig sites that cater to rockhounds, like the emerald mines of Appalachia, the diamond-studded dirt pits in Arkansas, the sodalite quarries in Ontario, and the thunderegg beds of Oregon. All over the continent, someone has opened up the ground and invited us in. Often these mines charge the public for digging, but the fees are usually modest. Visiting these mines provides an excellent opportunity to meet others who love the hobby.

Abandoned mines are a different beast altogether. North America is littered with sites of extraction that have been exhausted and deserted. Rockhounds love exploring these places for interesting material left behind. Many of these former mining operations are riddled with tunnels burrowing deep into the ground. Even a mine's entrance, or adit, is dangerous.

Adits are what most of us picture when we imagine an old mine—a horizontal tunnel supported by posts and beams that leads horizontally into a hillside. Abandoned mines are old places, blasted out with dynamite and supported with timber that's been dry-rotting for a hundred years or more. Under no circumstance is it ever safe to enter an abandoned mine. The Nevada Division of Minerals famously prints bumper stickers that read, "Abandoned Mines: Stay Out and Stay Alive!" as part of a larger campaign to address the tragic number of people who risk going in and never coming back out again.

The aboveground tailings are safer places to explore. These hill-sized piles are the overburden—or reject rock—that was removed from the mine to access the prize material. Often there will be interesting rocks and minerals littering the dumps (see page 49, bottom). Keep in mind that mine tailings are not exactly safe either. Certain elements can concentrate in the piles, making them a risky kind of dirt to mess around in. When searching through tailings, I like to wear gloves and, if it's dusty, a mask. If I can find a hose at the end of the day, I try to give myself and my gear a good wash down. The same goes for quarries, where rock has been pulled away from a hillside, leaving an exposed cliff with piles of tailings around the base. Old quarries carry another kind of risk. Think of them the way you would think of a sketchy roadcut in terms of stability. When the site was retired, no one said, "Hold on, let's stabilize this cliff before we disappear." Remember that before you take a hammer to it. Abandoned mines and quarries require your utmost attention and precaution. Always search aboveground. Avoid eroding cliff walls. Bring protective gear. If there are tunnels and adits around, there's only one rule to follow: stay out and stay alive!

While rockhounds must first learn to read the landscape to find places where rocks show up, we must also learn to read another kind of landscape: the one where laws, rules, and regulations apply to the land beneath our feet. These are the lines on the map that delineate not slopes or waterways but state and county lines, provinces and towns, and public lands and private properties. Though these boundaries and lines are invisible to the naked eye, they are very real. Especially when it comes to public versus private, it's downright necessary that rockhounds know the difference. This knowledge keeps us out of trouble (a lot of trouble) and also keeps the hobby-at-large in good standing. This is where map reading becomes an essential skill for rockhounds. It's our responsibility to deduce that we're in the right place and know what we're allowed to do there. Becoming familiar with the various kinds of land ownership may not be the most glamorous part of the learning curve, but it's definitely of utmost importance. Stay with me!

We can't begin discussing land ownership without acknowledging how these lands came into possession in the first place. Even though political and private land boundaries are a well-established social construct, they're not one that was ever mutually agreed upon. The United States and Canada came into being through colonization, war, and outright Indigenous genocide. These crimes were perpetrated by waves of immigrants from the European continent, and their descendants have kept a monopoly on political power and land ownership through the centuries. To this day, Indigenous peoples are still fighting to regain control of their homelands and access to resources that were taken from their ancestors.

Why does this matter to people who are just looking for rocks? Because rockhounding is a perpetuation of taking land, just in very small pieces. In the United States and Canada, there's no comfortable way to separate the hobby from the historic consequences of colonization. The seemingly infinite public lands that are open to rockhounds have come to us at a great cost. It's not the act of collecting rocks that's inherently compromised, though. Humans have always gathered rocks and minerals. It's part of our nature. It's purposeful. And joyful. The compromise comes from the privilege of collecting on land stolen from its original peoples. There's no way to ethically justify rockhounding in this context. That's

why it's important for each of us to exam-
ine our motives for collecting and what the
future holds for the material we take.

How do rockhounds strike a balance in
this situation? Is there a general wisdom
about treading lightly? Each person needs
to decide what that looks like for themselves. Collection rules on public lands
may allow you to take a lot of material,
but that doesn't necessarily mean that you
should. Ask yourself, if I take this rock from
the place, what is its fate? Will it remain
appreciated? Will I do something useful
or special with it? If the answer is yes, keep
it. If the answer is no, leave it. If the answer
is maybe, definitely leave it. It's the maybes
that tip the scale toward heavy-handed
collecting.

Rockhounding Rules in the United States

In the United States we are fortunate to
have a massive amount of public land.
Public lands are controlled collectively by

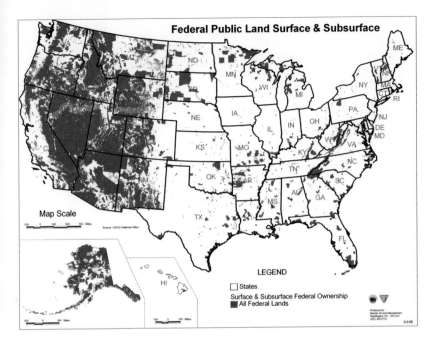

Federal Public Land Surface & Subsurface

LEGEND

☐ States

Surface & Subsurface Federal Ownership
■ All Federal Lands

Map Scale

Source: USGS National Atlas

US citizens and managed by government agencies. There are essentially three tiers of public land control in the United States: land managed by the federal government, land managed by state governments, and land managed by local governments of counties, cities, and towns.

All federally owned public lands are managed by the US Department of the Interior (DOI) and the US Department of Agriculture (USDA). Underneath the umbrella of these departments huddle numerous agencies responsible for managing public lands, and they each have different limits regarding rock, mineral, and fossil collection. The last major update of public land laws was passed under the Federal Land Policy and Management Act (FLPMA) of 1976. "Flipma," as you will hear it pronounced out loud, remains the legal framework from which all agencies issue rules relating to how we access and use federal public lands today. The act allows agencies to set their own collection limits, which is why rockhounds need to be familiar with each agency and its rules.

The most basic, simplified breakdown of what we can collect on federal land goes like this:

+ **ROCKS:** Yes!
+ **MINERALS AND METALS:** Yes!
+ **PLANT FOSSILS AND INVERTEBRATE FOSSILS (I.E., CORALS, MOLLUSKS, INSECTS):** Yes!
+ **VERTEBRATE FOSSILS (FOSSIL BONES AND TEETH):** No! A paleontology permit is required.
+ **ARTIFACTS:** No! An archaeology permit is required.

- **TRASH MORE THAN 100 YEARS OLD:**
  no! Old trash is considered archaeology.
- All collecting must be for personal use.

This basic set of rules remains consistent across all federal lands. It's the *amount* of material that we're allowed to collect that changes from agency to agency. Collecting is not necessarily allowed on all public lands, either. For example, rockhounding is forbidden in national parks and protected wilderness areas.

If you take anything away from this book, let it be this: always, always, always look up the collecting rules for the area you are intending to explore. Agency websites can be a good resource, but the higher path is to pick up the phone and call the agency office, especially if anything is unclear. The people who staff our public land agencies are tasked with the job of making this information available to us. I call public land headquarters regularly, and it helps me not only hear the rules firsthand, but also develop a rapport with the people who care for the places I love to explore.

Here's a general breakdown of the major public land agencies and the rockhounding rules for each. Please note that my overview is not the final word on collection rules, nor is it exhaustive of exemptions and small caveats that specific agencies may have. (That's why you always call the local headquarters first!)

BUREAU OF LAND MANAGEMENT

✢ ✢ ✢ ✢ ✢

The Bureau of Land Management (or BLM as most people refer to it) is the heavy hitter of public land managers in the United States. They manage one out of every ten acres of land in the United States; most of it is located in the West. Rockhounds are a big fan of BLM land since it has generous rules for rock and mineral collecting. On BLM land, rockhounds are allowed to collect up to 25 pounds of material per person, per day, with a yearly maximum of 250 pounds per person. Daily limits prevent people from taking giant things, and the yearly maximum works to prevent over-collection in general. Sometimes there are ways hobbyists can work with the BLM to collect beyond these limits. Say you find a large, petrified log that weighs much more than the daily limit. You can contact the local BLM district office and arrange a purchase agreement with them to remove it. The agreement has to be made *before* you remove anything oversized. However, my hope is that you'll consider leaving that petrified log in place for others to enjoy. Furthermore, if you plan to collect large quantities of rock for commercial purposes, you need to stake a mining claim (more on that to come!).

Occasionally, conservation designations pop up on BLM land to protect special places. These are wilderness areas, wilderness study areas (WSAs), and areas of critical environmental concern (ACECs). In almost all cases, rockhounding is limited or prohibited in designated wilderness areas and ACECs. These sensitive places are well marked on maps. Keep your eye on those boundaries and you'll keep out of trouble.

Some national monuments are also managed by the BLM. If so, rockhounding

may be allowed. However, if a national monument or similar designation is overseen by the National Park Service or US Forest Service, rockhounding is off limits. Designations on par with national monuments are national scenic areas, national recreation areas, national conservation areas, and national grasslands. Always check the rules in these special conservation areas. Make a concerted effort to learn which agency oversees them and what the rules are.

NATIONAL PARK SERVICE

✦✦✦✦✦

This one's easy. If it's a national park, you cannot rockhound. Hands off, my friends! The National Park Service is charged with protecting these places for the visual pleasure of humanity and the conservation of its flora and fauna. You've probably visited a national park—and perhaps you've taken a rock or two while there. It's OK. Just don't do it anymore. Violators of this law are subject to criminal penalties and the National Park Service does indeed enforce the law.

One place where rock collecting has become a serious problem is in the Petrified Forest National Park in Arizona's Painted Desert, one of the most stunning fossil wood locations on earth (see page 60). My wife, Lisa, and I did a monthlong artist residency in the park some years ago. During our time there, park rangers regaled us with stories about visitors who stole pieces of petrified wood from the trails, only to toss them out of their cars and RVs just yards from the park exit. People would panic when they saw signs stating that all vehicles were subject to search. After the gates had closed for the evening, rangers would walk that stretch of road to collect the sparkling pieces of petrified wood people had thrown out of their vehicles.

The most tragic aspect of this endless relay of rocks is that park rangers can't put them back. They never know what part of the park they have been stolen from and guesswork isn't in the rulebook. Instead, rangers take these beautiful specimens and toss them onto something called the *conscience pile*. I saw the pile with my own eyes once. I wanted to weep. People have really good taste. They steal particularly beautiful rocks. And here they were, in a lonesome mound on the side of a service road, tucked out of sight. It was the most magnificent, colorful, sparkling collection of petrified wood I had ever seen in my life. Imagine if these beauties were still along the trails where people could enjoy them! Some of the specimens in the conscience pile had actually made it past the gates at one point. These pieces joined the fated pile after being mailed back to the park by regretful visitors, often with a "conscience letter" explaining how the stolen rock had cursed them. Park rangers keep a collection of these letters in a display case in the visitor center, hoping it will compel people to think twice about stealing. If you're curious about what kind of bad luck these ancient wood chunks are capable of bestowing, read *Bad Luck, Hot Rocks* by Ryan Thompson and Phil Orr. Their gorgeous book features photographs of rocks from the conscience pile as well as copies of the sobering letters that were mailed back to the park with them.

Now that I've convinced you never to take material from national parks, please know that there are a few exceptions. Some national parks in Alaska allow limited collecting, and a small area of park service land in Northern California (the Whiskeytown National Recreation Area) allows gold panning. The National Park Service also manages national seashores, which allow beachcombing. Most national seashores limit activity to a gallon of material. However, it's imperative that you look up the rules for the specific beach you intend to visit, since the guidance may vary.

USDA FOREST SERVICE

✣ ✣ ✣ ✣ ✣

The US Forest Service is another familiar public land agency and well established in all 50 states. It's likely you've spent time in national forests, which the Forest Service manages on behalf of the public as a timber commodity. Forests may not be the first place you'd consider when looking for rocks, but these places often host rocky

outcrops on slopes and cobble-filled rivers and streams. They can be a favorite among rockhounds. The federal guidelines for collecting on Forest Service land are similar to those of the BLM, except the weight limit is lower. "Generally, a reasonable amount is up to ten pounds," explain most Forest Service brochures. A permit is required to dig or use heavy equipment on Forest Service land, as well as for using sluice boxes for gold prospecting in rivers. If you plan to collect large quantities of rock for commercial purposes, you need to stake a mining claim. Always contact your local headquarters for their specific guidance. There may be some sensitive tracts of forest where collecting is off-limits. Make sure not to confuse national forests for state forests, either. State forests may have very different rules. There's plenty of private land within forests, often owned by timber companies. Avoid private forest land at all costs!

US FISH & WILDLIFE SERVICE

✦✦✦✦✦

The US Fish & Wildlife Service is another agency that has rules that can change from location to location. Fish & Wildlife's mandate is to protect hunting and fishing stocks within national wildlife refuges, and rock and mineral resources are not always in their purview. In many scenarios, it's possible that rockhounding is not allowed within certain wildlife refuges. This is where my insistence about inquiring with the local headquarters is key.

Rockhounding on US Fish & Wildlife land is not nearly as straightforward as it is on BLM or national forest land. For example, in a remote corner of northwest Nevada, US Fish & Wildlife manages the remnants of an ancient volcano known as the Virgin Valley caldera within the Sheldon National Antelope Refuge. Geothermal waters bubble up from the collapsed caldera rim, filling the basin with year-round springs brimming with life, especially bird species. The caldera also happens to be one of the finest fire opal mining locations in

North America. To keep the impact low, US Fish & Wildlife limits rock collecting on the refuge to seven pounds per person, per day. However, miners have found a way to get around those limits by staking mining claims. (Yes, I am one of them. We'll get to that in a minute because mining claims are a wild, wild departure from the "tread lightly" ethos I want to distill in your heart!) To prevent the spread of mining into sensitive wildlife areas, US Fish & Wildlife has closed off large sections of the refuge to miners. It helps to contain the madness, but between the checkerboard of mining claims and off-limits wildlife areas, it can be a very confusing place to rockhound, especially for people who don't have claims.

NAVIGATING MINING CLAIMS ON PUBLIC LAND

✦ ✦ ✦ ✦ ✦

Eventually in your rockhounding adventures, you'll come across mining claims on public land. You'll wander into an area with many unofficial-looking posts sticking out of the ground. These posts mark the corners of mining claims—parcels of public land in which someone has exclusive rights to specific rocks and minerals within it. Shortly after the Civil War, a suite of federal laws was passed that allowed anyone to stake a claim for mineral rights on public land. Known in the rule books as the General Mining Act of 1872, it spurred the Gold Rush and sweeping westward expansion. Surprisingly very little has changed about the mining laws of 1872 since their passage. Today, both small-time miners like myself and massive mining corporations operate on simple, inexpensive mining claims acquired through these antiquated laws.

How does staking a claim work? Imagine you've come across a deposit or vein of something cool on public land—garnets! turquoise! opal!—and you want to collect it in abundance and perhaps sell it commercially. No matter which agency manages the land you found it on, you'll approach the BLM specifically to file a mining claim. By staking a claim, you are withdrawing the mineral rights from public use and reserving them exclusively for yourself. You can keep the claim indefinitely if you file the correct paperwork and pay a small fee each year. Existing mining claims can also be bought and sold on the private market, but the BLM is the agency that keeps tabs on all mining claims under the legal framework set forth in 1872. The most important aspect of a mining claim is that *the land remains public* even if you own the mineral rights. Anyone can hike through or camp on your claim, they just can't collect claimed rocks on it. If you come across mining claims on public land, there are some basic facts that can help you assess their boundaries. All mining claims are a standard size, approximately 20 acres, and always rectangular. If a discovery spans an area larger than 20 acres, miners can stake multiple claims, but each 20-acre parcel must be marked individually. Each corner is marked with a post, e.g., the northwest corner needs to read "NW corner," and so forth. A fifth post, known as a landmark, is sometimes placed in the center (see opposite page). The landmark post will have a tube or jar attached with the claim owner's notice inside. Thorough miners will

even leave a map. If you wander into an area with mining claims, the best thing to do is read the marker on the closest post to figure out where the 20-acre rectangle sits in the landscape. From there you can assess the boundaries and stay outside of the claim.

How do I balance my own "tread lightly" ethos with the fact that I also have mining claims? Ironically, I staked my claims in the Virgin Valley caldera to protect an area, rather than to mine it out. I had come across an area rich in petrified wood, with fire opal–filled tree stumps poking out of the ground. The area had been claimed in the past, and the previous miners left open pits and trenches everywhere. They must have hauled out some extraordinary logs. I knew it was likely this area would be claimed again, and I was sad about what that might mean. So I staked two 20-acre claims to save it from a harsher fate. Yes, staking a claim has major colonizer vibes. There are aspects about it that feel good, though. I can bring my guided tours there and let them explore to their heart's content. I can let them collect without stripping the area of its treasure or ruining the remaining petrified stumps. I'm working on filling in the old trenches and pits, and making it look less forsaken. It's my experiment with a strange and outdated law.

STATE-, COUNTY- & LOCALLY OWNED PUBLIC LAND

State forests, beaches, parks, and other types of local public lands can be excellent rockhounding locations, especially in the central and eastern part of the United States. However, this is where rockhounding rules start to become harder to categorize because they're not unified the way federal rules are. And it's why my steadfast belief in calling your state or county land office for information will remain good advice forever. Many of these places are set aside for conservation efforts, but you'd be surprised that some don't have any guidance concerning rockhounding. Usually, rockhounds can interpret a lack of prohibitions as permission to collect. However, we need to do our research before assuming this (never assume!). Going to your local state forest to check out a rocky outcrop in the woods? Call the state forest office. Want to collect pebbles with your students from the stream in the county park? Call the county park office first. Once you find out the rules, you can encourage the agency to post them publicly and make the world an easier place to rockhound.

PRIVATE LAND

✤✤✤✤✤

Unlike public lands, private properties are owned by individuals and can be sold as a commodity on the market. If land is privately owned, it means you are not allowed to enter without the landowner's permission. This is one of the most important distinctions in rockhounding. Unless you are invited by the owner, you are forbidden from collecting on their property. However, if you are invited, private land affords many rockhounds opportunities to collect beyond the limits set on public lands. Personally, I find this idea unsavory. Other rockhounds consider it a dream!

Private landowners are afforded greater rights of use compared to the limiting laws that apply to public lands. If you are a private landowner in the United States, you are free to remove almost anything from your property: rocks and minerals, any sort of fossil, artifacts, and so on. Found a *T. rex* on your ranch? You can dig it up! Have the ruins of a paleolithic pit house in your meadow? Yes, you can excavate it! Occasionally limits do apply. For example, the government may include some of your property in artifact conservation laws, however this is rare. More likely, you may own a particular piece of property, but you may not own the mineral rights to the ground beneath your feet. If this is the case, removal of rock material is reserved solely for the owner of the mineral rights. It's unusual for this to be the case, but not unheard of. In an arid and lonely basin in the heart of the Mojave Desert, there's an island of private property within an ocean of public land. Within that island, I own a small five-acre parcel, but I don't own the mineral rights. Ironic that a rockhound would own land that they can't collect rocks on, don't you think?

INDIGENOUS NATIONS & RESERVATIONS

✢✢✢✢✢

Truth: all land is Indigenous land. Until we can all collectively dive headlong into the #landback movement (something I'm working on with my own property), the governments of the United States and Canada only recognize small swaths of land as legally belonging to native peoples. These areas are Indigenous nations and reservations, and they are sovereign lands, meaning the people have their own government and powers to regulate their internal affairs. Unless you are a member of a sovereign tribe with permission to take from the land, these nations and reservations must be treated by rockhounds like all other private property. Without explicit permission, collecting is forbidden. Take nothing.

Rockhounding Rules in Canada

Rockhounding rules in Canada vary across the country, with each of the ten Canadian provinces and three territories establishing general mining laws that characterize hobby collecting, or rockhounding, as a privilege, not a right. Hobby collecting is considered separate from all other mining activity. This aspect of Canadian rockhounding guidance can make it a little bit more of a research project to suss out rockhounding rules in the area you'd like to collect in.

Public lands in Canada are known as *crown lands* and collecting is allowed on a great majority of them. Rockhounds must first ascertain that the mineral rights *and* the surface rights are owned by the Canadian government before rockhounding. This is usually the case, but it's important to do your research. There are special designations of crown lands that prohibit rockhounding for conservation or safety purposes such as provincial and federal parks, First Nations reserves, abandoned mines that have been posted as a mining hazard, and areas freshly leased for industrial mining. Similar to the United States, some crown lands can be withdrawn from public access for other conservation efforts. The Ministry of Natural Resources is an excellent resource for rockhounding guidance, as are district geologists who work for the Geological Survey of Canada. Consultants in provincial and territorial Geological Survey division offices can consult rockhounds on legal questions and mining claim inquiries.

CHAPTER 2

Every Rock Is a Story

"To seek and save rocks is to seek and save stories of the earth."

—Brooke Eberle, geological engineer and rockhound in Colorado

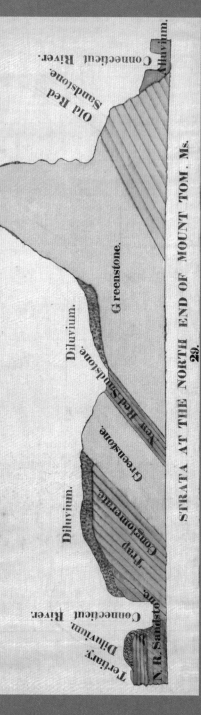

STRATA AT THE NORTH END OF MOUNT TOM. Ms.

29.

Connecticut River.

Old Red Sandstone.

Alluvium.

Greenstone.

Diluvium.

Ver. Red Sandstone.

Diluvium.

Greenstone.

Trap Conglomerate.

N. R. Sandstone.

Connecticut River.

Tertiary Diluvium.

A rock is a moment frozen in time. Its story is never boring. No matter where you are in North America, the rock material that makes up the ground below you tells the ever-evolving story of our planet, connecting us to the past and projecting us into possible futures. Depending on where you live, the rocks under your home could have been born in an explosive volcanic eruption or cemented over the eons deep on the ocean floor. They could have washed down an ancient mountain range or built up along a coral reef. In the millennia between their creation and our lives, many of them have shifted, tumbled, crushed, and recrystallized. The Earth's crust is a book of memories. A compilation of its past lives.

For rockhounds, it's what's at the surface that matters. As you cross the continent, the types of rock material change dramatically. Parts of our continent have been sitting around since the early days of Earth, while other regions are new additions, glued onto the continent as incoming ocean plates dip beneath our own, tossing bits of crust our way. In fact, the entire

West Coast is constructed of these glued-on bits, making for a fascinating variation in surface geology. Learning to read the ground underneath your feet is a key aspect of successful rockhounding. It also makes the hobby so much more interesting when you can look at past a rock's stripes or speckles and see millions of years of history unfold before your eyes.

Many rockhounds set out in search of specific rocks or minerals. Some of the most popular materials to collect are agates and jaspers, thundereggs and geodes, petrified wood, marine fossils, and crystals. All these materials are common in North America, but they don't show up everywhere. They only show up in specific geological situations. If a rockhound is looking for something specific, it's their job to seek out the specific geological situation in which the material occurs. This requires a willingness to look up geological information and make sense of it. A good rockhound is a good researcher.

I'm not a geologist, so I'm not qualified to provide a serious introduction to geology.

But I am a professional rock-finder, and a deeply nerdy geology enthusiast, so I can give you advice on how to approach the science from an amateur's perspective and use it to your advantage. I can share with you my understanding of the science and promise you that there are ways to work through the dense jungle of information and come out unscathed! In your journey to become an intrepid rockhound you're going to have to jump right into the foreign language of geology. You're going to encounter a tremendous amount of vocabulary that's new to you. There are a few things non-geologists like us can do to make sense of complex information. Here's my best advice:

✣ **BECOME FAMILIAR WITH THE GEOLOGIC TIME SCALE.**

There's no need to memorize the geologic time scale but purchasing some sort of reference can be super helpful. Being familiar with it matters since almost all rock formation discussions reference the era in which they formed. Consider picking up an illustrated time scale poster or booklet. I have a favorite coffee mug that has the geologic time scale on it, so I get to study it every morning.

✤ **FEEL COMFORTABLE IN THE ROCK CYCLE.**
Remember learning about igneous, metamorphic, and sedimentary rocks? Often these are taught as rock "types," but in truth it's way cooler to think about them as processes: How did these rocks get here? They blew up in a volcanic eruption (igneous). Or they settled here in a valley and cemented (sedimentary). Or they baked deep within the crust, slowly changing from their original form (metamorphic).

✤ **DON'T BE INTIMIDATED BY BIG WORDS.**
Geology is a language of complex terminology. Don't understand a term? Just look it up! Often that definition will lead to something just as dense, so continue to break down the answer until it reads plainly. Wikipedia is one of my favorite references for breaking down geological terms.

✣ **BE WILLING TO THINK ABOUT THE ELEMENTS.** Remember the periodic table? Well, only a handful of those elements are truly abundant on Earth. Oxygen, silicon, iron, and aluminum take up the lion's share—about 80 percent of the crust. So, if you're picking up rocks, it's likely that they're largely composed of those top four elements. Calcium, sodium, magnesium, and potassium follow in abundance, with all other elements scattered quite sparsely throughout the crust.

✣ **DON'T TAKE JUST ANYONE'S WORD FOR IT.** The best resource on the geological history of a place is always written by a geologist. Information about rock formations on rockhounding forums and blogs can often be inaccurate. No one is trying to lie. It's more that the writer fails to research geological information in depth. Instead, they repeat whatever information they've heard and move on. *Experienced rockhound does not equal geologist* (and that includes me). Furthermore, rockhounds famously make up their own names for rocks that have little to do with actual geology. Take rockhound jargon with a grain of salt.

Finding the Goods

The big questions for people new to the hobby are *what* shows up and *where*. It's not the quickest question to answer when you're talking about an entire continent, but let's give it our best shot. We'll take the rock-cycle approach since it feels the most universally understood. In each rock creation process, very specific rocks and minerals appear, so it's perhaps the best way to narrow down what kinds of rocks show up and where.

ROCKS FROM THE DEEP

Metamorphic rocks are visitors from the deep. Consider them the basement of the Earth's crust exhumed and floating at the surface. These rocks make it all the way up under specific circumstances: either the

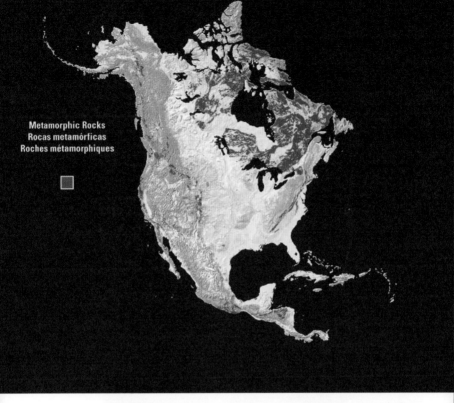

Metamorphic Rocks
Rocas metamórficas
Roches métamorphiques

earth on top of them entirely eroded or they were forced up in mountain-building events. Sometimes the squeezing from mountain building alone is enough to bake and recrystallize the upthrusting rock.

In this map you'll notice a large island of metamorphic material surrounding Hudson Bay in Canada. This is the Canadian Shield, some of the oldest rock material in North America. You'll also notice skirts of metamorphic rocks following the flow of the Appalachians, Rockies, and California coast ranges. These are exposures associated with mountain-building events known to geologists as *orogenies*.

Slow baking deep inside the earth allows large crystals to grow, like the garnet, pictured on the previous page. The pressures often squeeze crystals, both visible and microscopic, into waves and bands, resulting in the layered schists and wavy gneisses we see in some mountain ranges today. Sedimentary materials that form at the surface, like sandstones and limestones, can be forced deep into the crust where they also bake. Under heat and pressure they recrystallize, forming quartz-ites and marbles that are sometimes forced back to the surface. (Lucky us! Especially the sculptors!) The most important aspect of metamorphic materials is that they get so hot they bake—but not so hot that they melt. Once melting occurs, we enter the world of igneous rocks.

The other category of deep earth material is the plutonic rocks. Plutonic rocks are associated with melting, so they're igneous, and the materials they produce are crystals that emerge from the melt as it slowly starts to cool. In many cases a pluton can melt and cool multiple times, most famously resulting in the formation of granite deep underground. Sometimes this plutonic material makes it to the surface through eruptions. See a volcano? There's a pluton underneath! Plutons are the sources of magmas (melts within the crust) and lavas (melts that explode onto the surface). If a large region is volcanically active, you can assume there is a large pluton underneath. As volcanoes erode over millions of years, the pluton may rise

to the surface, forming what geologists call a *batholith*. There's a good chance you've walked among these former basements of long-gone volcanoes: the White Mountains of New Hampshire, the Sierra Nevada of California, the Coast Range Arc of British Columbia, the Bitterroot Mountains in Idaho and Montana, the monzogranite boulders of Joshua Tree, and more.

So what's good for rockhounds in these deep earth rocks? The key is to think *crystalline*. Since the processes of baking and melting are slow, this is where crystals show up. Almost all precious gemstones come from areas where high temperatures and pressures have formed large, gorgeous crystals. Rockhounds search metamorphic and plutonic rocks for veins and clusters of well-known crystals such as rubies, sapphires, garnets, tourmalines, quartz points, and so on. If precious gemstones (crystals with value) are your thing, then searching in metamorphic and plutonic rock formations is likely to bring you luck. Even more so, it's where these hot rocks and plutons rub up against other types of rock formations that interesting things happen. In these places something called *pegmatites* can form. Pegmatites are veins of especially large crystals, so keep this term in your pocket as you search for the best places to hunt.

Plutons are an important source of valuable metals as well. Metals tend to concentrate along the margins and in veins of plutons as ores (metals blended with other elements), especially if hydrothermal activity (groundwater mixing with hot rock) has occurred (see page 92). Rockhounds who search exclusively for valuable metals like gold, silver, and copper are known as prospectors. Finding native metals (pure nuggets not bonded with other elements) is the fever dream of all prospectors. Deposits of copper ores yield some of the most famous gem material too: azurite, malachite, turquoise, and chrysocolla. Metal prospecting can become a lucrative hobby but requires a willingness to learn some geochemistry and excellent wayfinding skills, as well as strong arms for a whole lot of digging.

Volcanic Rocks
Rocas volcánicas
Roches volcaniques

ROCKS THAT ERUPTED

✢ ✢ ✢ ✢ ✢

If plutonic material can make its way from deep in the crust to the surface of the earth, it makes its way there volcanically. Typically, when people think of volcanoes, they think of a cone-shaped mountain exploding with red, hot lava. In truth, this is only one type of eruptive body, and

it's not necessarily accurate either. It's an image simplified in the popular imagination. In reality, molten rock makes its way to the surface through a variety of means. It can come up through mere cracks in the ground, flooding a region in a blanket of lava. This is how most of Oregon and Washington came to be covered in layer after layer of dark lava rock. Or it can explode so powerfully from a metropolis-sized caldera that it blankets an entire continent in dusty, powdery ash, causing years of global cooling. This is the type of eruptive body underlying Yellowstone National Park. On the map we see some ancient volcanism dispersed in the northern stretches of Canada around the Hudson Bay, but the great majority of volcanic rocks are associated with mountain ranges. Much of the west coast of North America is made of volcanic material. This is due to all those island chains glomming onto the continent as Pacific Ocean plates slide underneath North America. That sliding is known to geologists as *subduction*, which causes all sorts of melting as the ocean plate slips under, leading to volcanoes popping up as those melts rise to the surface.

Why do rockhounds need to know all of this? Because different kinds of volcanoes spew different kinds of material. The style of eruption (and the shape of the volcano) is entirely defined by the chemistry of the pluton beneath it. This is where those top four crust elements come into play: oxygen, silicon, iron, and aluminum. Plutons richer in iron and aluminum create lavas that run fast and hot; think of those quick-flowing molten rivers of Hawaii. However, as the balance shifts toward more richness in silica (silicon and oxygen),

magma in the pluton gets thicker, steamier, and more explosive because more force is required to move it. When it erupts as lava, it doesn't flow nicely. That's when tall volcanoes start to build up above the pluton. These cone-shaped volcanoes are just giant piles of sticky lava that didn't flow very far. Consider the beauties of the Cascade Range in the Pacific Northwest: Mount Rainier, Mount St. Helens, Mount Hood, Mount Shasta. These volcanoes look the most like the ones in our popular imagination, but their explosions are much ashier and debris filled. Yes, some red lava flows, too, but not as much as you'd think.

Sometimes the silica content in a pluton can become so high that things get crazy. This is typical of large hot spots under continents. The pluton melts and cools and melts and cools and because of the way chemistry works in the crust, silica just keeps concentrating. This makes for an almost immovable magma. Eventually, the pressure will build up so intensely that the entire earth above the pluton explodes and collapses. These are our hot spot calderas, or "supervolcanoes," that disaster TV shows like to get so hyped up about. The Yellowstone hotspot is an active supervolcano with a long history of eruptions that began about 17 million years ago. The surface of the Earth moves around a lot. When eruptions from this hot spot began, the earth above it was the Oregon-Nevada border. Over time, the continent drifted over the hot spot. If you look at a satellite image of Idaho, you'll see a broad curving plain, known as the Snake River Plain, that follows the hot spot's path of eruptions over the last 17 million years, connecting Oregon to Wyoming. Currently the hot spot sits under Yellowstone National Park, heating the crust so intensely that geysers and thermal pools riddle the landscape. Yellowstone will eventually blow again, but it is not likely to happen in our short lifetime.

For some folks all this chemistry might be dull, but it matters a lot to rockhounds. Different lava chemistries yield different treasures. High-iron lava yields mostly basalts. Basalts are those gorgeous, black lavas, sometimes filled with gas pockets.

Basalt flows leave very cool lava tubes behind that are incredible to explore, like long, winding caves. In the gas pockets of basalts, unusual minerals can form—namely zeolites, which are prized by collectors and which also have some industrial value. Agates and jaspers can also form. In the deserts of southeast Oregon, some of the ancient flood basalts yield large, beautiful feldspar crystals known as sunstones (see page 101).

But it's when lava starts to get higher in silica that the real treasure hunt begins. High-silica lavas give us the most famous rockhounding materials: agates (see above left), jaspers, geodes, thundereggs, opals, and petrified wood. These materials form as all that silica goodness (silicon and oxygen) in volcanic ash dissolves in groundwater after eruptions. The dissolved silica eventually redeposits (precipitates) in bands, veins, layers, gas pockets, and organic matter (in some cases fallen trees) to form an endless variety of treasures. Silica is a very hard mineral, so silicate rocks like agate and jasper make excellent lapidary materials. Obsidian is also a beloved volcanic treasure (see opposite right). It's volcanic glass, material that cooled so fast none of the atoms within it had time to arrange into crystalline patterns—a frozen chunk of chaos. It, too, makes an excellent lapidary material, and rare rainbow varieties are highly prized by collectors. High-silica volcanic lavas and ashes can also cool into a stunning array of rhyolites and tuffs with intricate orbicular and banded patterns—something I seek out regularly to use in my craft. "Picture jasper" is one of the most famous rocks that is born of cooled volcanic ash.

There seems to be a lot of confusion in the rockhounding community about how silicate treasures like agates, jaspers, opals, and petrified wood form, so let's break it down. Volcanic rocks and ash blanket the ground after big eruptions. Over time, rocks and ash build up in layers as eruptions continue. Groundwater travels through these layers, dissolving some of the silica, especially from the ash. As the groundwater travels around, the dissolved silica within it precipitates in voids, seams, and gas pockets, and invades organic material—from a single fallen tree to entire smothered swamps. Agates, jaspers, opals, and petrified wood are manifestations of silica that has dissolved in groundwater and redeposited. Agates and jaspers are the same thing, composed of microcrystalline to cryptocrystalline quartz (quartz is pure silica). Quartz crystals, by contrast, are macrocrystalline. The prefixes refer to crystal size. In agates and jaspers, the crystals are infinitesimally tiny. *Agate* and *jasper* are just rockhounding names. In geology, these kinds of deposited silica are referred to as chalcedony. In the rockhounding world, agates tend to have some transparency while jaspers are opaque. That's the only difference. Inclusions of other minerals lend color to the stones. Both agates and jaspers can have orbs, bands, mosslike patterns, and pockets of druzy (sparkly macrocrystalline quartz). Thundereggs and geodes are voids in volcanic rock and ash beds that have only partially filled with quartz, opal, agate, and jasper. In Oregon, where I live, thundereggs (silica-filled gas pockets) are the state rock, and digging for them is a favorite pastime for many (see page 104).

Opals (see opposite page, top) form the same way as agates and jaspers—as a silica deposit—however they still have some of that groundwater in their crystal matrix, so opal is known to geologists as *hydrated silica*. The water within their crystal matrix makes opals terribly fragile and temperamental to work with. When exposed to air for the first time, they can crack and crumble. Fun! Opals can be transparent or totally opaque, have fire (iridescence), have schiller (a flash), or have mosslike inclusions. They can bring in some very big dollars if they're flashy enough. Sometimes fossils can be found that are replaced entirely by precious opal, especially petrified wood.

Petrified wood (see opposite page, bottom left and right) is my favorite material to search for, and you may remember that I have two fossil wood–mining claims in Nevada. The trees on my claims were buried in the massive crater lake that formed after one of the very first Yellowstone hot spot eruptions—back when it sat under the Oregon-Nevada border about 16 to 17 million years ago. The massive petrified tree stump pictured at left was preserved in such an explosion. It's located a few dozen miles south of my own claims. Because of its enormity, the Bureau of Land Management fenced it in to protect it from vandalism.

Petrified wood and other petrified organic materials fossilize as silica-rich groundwater invades their cellular structure. These are trees, twigs, pine cones, soils, and swamp beds buried in ash after a volcanic explosion. So how do some of these things, especially wood, fossilize so perfectly? The tree's internal structures that facilitate the movement of water up through the trunk while living still function passively after burial in sediments. They love to suck up water. Some high-school chemistry for you: the tree's woody structures (lignin and cellulose) tend to become acidic when invaded by groundwater, and silica dissolved in water tends to be alkaline. (Remember acids and bases?) When the two substances interact, a pH potential is created, causing the silica to replace the carbon structures that make up the wood instantaneously. The molecules swap places. Bam! This is how wood petrifies so beautifully. Silica literally and microscopically replaces wood on a molecular level.

The first type of silica to precipitate in wood is usually opal. The petrified wood samples in the righthand part of the photo are "opalized." Over time the opal can dehydrate, shedding its water and transforming into chalcedony (agate and jasper). Rockhounds would call this type of petrified wood *agatized*. Both opalized and agatized wood are common in volcanic sediments because eruptions tend to knock down trees, and buried trees love to suck up ashy groundwater. If you want to get extra nerdy about petrified wood formation, look up the research of George Mustoe. His scientific papers are easy to read and full of amazing images. He loves to share his sources for finding wood, especially in western states.

ROCKS THAT STACKED UP

✦✦✦✦✦

If you're sentimental about sediments, boy, are you in luck! Just look at that map (see opposite page). You'll notice that almost all of North America is blanketed in sedimentary rock. Anything that erodes from its host source and piles up somewhere else is sediment. Sedimentary deposit types can range from ancient, hardened rock to the cobbles that line the creek behind your house. Some of the most famous sediments in North America make up the stunning sandstone canyons of Utah, the lonesome badlands of South Dakota, and the limestone sinkholes that terrorize the populace of Florida. Large swaths of the Midwest are ancient marine sediments from a time long before the dinosaurs. If you live in a valley, you are living on sediments from disappearing mountains. If you're alligator hunting in a bayou, you're boating over river-delta sediments. If you're visiting my mining claim in Nevada, you're digging through volcanic sediments. To quote Charles Darwin, sedimentary layers come in "endless forms most beautiful."

Sedimentary Rocks
Rocas sedimentarias
Roches sédimentaires

What's good for rockhounds in sedimentary layers? Fossils! Often the best terrestrial fossils are preserved in either ashfall deposits from volcanoes (a sedimentary rock with an igneous source), or swamp and riverbed layers. Marine fossils are best preserved in limestones and mudstones. Both environmental scenarios tend to fossilize vertebrate and invertebrate fossils quite well. In most cases, fossilization

of shells and bones occurs the same way petrified wood forms: dissolved silica in the groundwater invades the organic material and replaces it. Occasionally silica is not the preserving material. In the Appalachians and parts of the West, a wealth of fossil material is preserved in carbon-rich coal beds. Sometimes fossils are not preserved by other minerals at all. Often in fossil leaf localities, the material is still composed of its original carbon molecules. I guess you could consider them merely mummified!

Not all sedimentary layers are created equal when it comes to fossil resources. The best way to find rich deposits is to consult guidebooks and forums. One of my favorite guides is by Albert B. Dickas, which covers fossil sites in each of the lower 48 states. I also love to look up scientific papers on fossil sites written by paleontologists. (I use scholar.google.com to search for reports.) The most import-ant thing to remember about fossil col-lecting is the rule concerning vertebrate fossils on public land. Unless you have a paleontology permit, collecting fossil bones is forbidden. However, fossil plants, petrified wood, and invertebrates are legal to collect.

Some rockhounds are after the sedimentary rocks themselves. I love to search through sandstones and limestones for material I can use for sculpture. Other rockhounds love to search for strangely shaped iron concretions that can be found among sandstones (see page 112, top right). Crystal collectors often search sedimentary layers of ancient lakebeds that dried up thousands to millions of years ago for gypsum, selenite, and halite crystals (see page 112, top left). There is a rising tide of artists who seek out colorful clays and paleosols (ancient soils) to use as natural pigments and homemade paints. Agates, jaspers, and geodes can also occur in some sedimentary formations if there's enough circulation and deposition of silica happening over time. Many folks refer to jaspers that form in limestones, mudstones, and sandstones as flint or chert. Chert can also more specifically refer to deep-sea mudstones that

were built up by the silica skeletons of radi-
olarians (they're exquisite—look them up!)
as they sank to the seafloor over the eons.

If you're wanting to learn more about
the fascinating world of sediments, I highly
recommend Siim Sepp's Sandatlas
(sandatlas.org). It's a wonderful blog full
of easy-to-understand information and
beautiful images. Another excellent online
resource is Brian Rickett's *Geological
Digressions* (geological-digressions.com).
Brian's blog branches out into just about
everything related to geology, and his
articles are wonderful for people like us—
the non-geologists.

CHAPTER 3

A Good Rockhound Is a Good Researcher

"It's a good thing for us that the old-time prospectors did so much footwork. There isn't much that has escaped their notice, and it's all written down somewhere. You just have to find it! Save your knees and do a little research first."

—Martin Holden,
geologist and rockhound in Washington

GEOLOGICAL MAP
OF THE
UNITED STATES
COMPILED BY
C. H. HITCHCOCK, AND W. P. BLAKE
from sources mentioned in the text.
1874.

Rockhounding isn't all about luck. Sure, there have been times when I've pulled over on the side of a dirt road, or gone for a dip in a river, and found beautiful rocks, but my greatest finds have always occurred when I've done research ahead of time. Our hobby is wonderfully accessible if you're willing to get a little bit nerdy. The research skills required for rock-finding success are all about *who*, *what*, and *where*. You'll want to become comfortable with finding out who owns or manages lands; determining what types of rocks might show up there by consulting guidebooks or geological maps; and navigating your way in and out of collecting areas with confidence (and without cell phone reception). The more time you spend developing these skills, the more your repertoire of interesting rock material will grow.

For people who are starting out in the hobby, the two big questions are always the same: "Where do I go?" and "What will I find there?" The best place to start is by acquiring rockhounding guidebooks that cover collecting sites within your state or region. Rockhounding guides list popular collecting sites with helpful directions and waypoints. Falcon Guides and the Gem Trails book series are two popular resources published coast to coast. Even dusty old guides published many decades ago can lead you to interesting places with cool material. The important thing to know is that *no rockhounding guide is perfect*. Some sites listed in these books may no longer be accessible, and sometimes land ownership and access rules have changed after a guide has been recently published. A good way to think about rockhounding guides is that they are a collection of suggested places that you should still research yourself. Sometimes this means calling agencies ahead of time or keeping an eye out for mining claim posts when you arrive in an area. I'm particularly careful about public forests since timber operations can show up seemingly overnight and dramatically change accessibility. If a guidebook publishes information about a collecting site on private land, I always get in touch with the landowner and get verbal permission beforehand. This advice goes for information published on rockhounding forums and blogs as well. Always confirm the accessibility and legal land status of the place before you go. There's plenty of misinformation on the web.

I love rockhounding guides and still use them a decade into my hobby, especially when exploring areas new to me. However, I've found that the Roadside Geology book series is an even better resource for what interests me. These geology guides are not rockhounding guides—they don't focus on collection; rather, they focus on appreciation. Roadside Geology books are published for almost every Canadian province and the majority of US states. They do a wonderful job of explaining what types of rocks are underfoot, how old they are, and how they got there. If you're curious about geological time, don't go rockhounding without one of these wonderful guides!

In regions of the West where I feel at home, I've transitioned into a deeper mode of research for locating interesting

rock material. I take my queries directly to large, colorful sheets of paper that tell me all about the rocks underfoot in exquisite detail: geological maps. These are maps compiled by geological surveys—groups of people employed by the governments of Canada and the United States over the last two centuries to map the bedrock of the continent. The geological maps we have today are built on an almost incomprehensible amount of human labor; generations of scientists surveying the landscape from mountain to valley through scorching summers and frigid winters, taking rock samples, and outlining the delicate boundaries of rock formations along the way. Almost all topographic and geological information about North America was compiled by people traversing the landscape one step at a time—before a single satellite ever turned its gaze back onto Earth.

Something I've come to appreciate over the years is the density of information that can be found on a geological map—an epic story told in a single image. Geological maps hold an astonishing amount of information: rock types and ages, slopes and waterways, earthquake faults and landslides, mineral deposits and mine sites. They're also large documents with lots of tiny text, so I've learned that there's no better way to read them than printed on a giant sheet of paper. You can roll them out on your kitchen table and write notes all over them. I've even laminated the few that I repeatedly take out into the field. Reading geological maps, however, requires a little bit of a learning curve, but it's a fun one. The most important thing to know is that a good map will teach you how to read it.

Geological maps are always available to the public. In Canada, you can acquire maps through the Geological Survey of Canada (GSC). In the United States, the best resource is the United States Geological Survey (USGS), and their National Geologic Map Database is the best place to access geological maps. In addition to the USGS (a federal agency), each state has its own distinct geological survey, and geological maps can also be found through state agency databases as well. It's important to know that the names of these agencies vary greatly from state to state. Some state agencies call themselves "geological surveys," but others call themselves "bureau of mines," or "soil surveys." The title is usually based on the most common geological need the state has. Some state surveys publish rockhounding maps and fun articles about where to find rocks. The state geological surveys of Utah and Wyoming are particularly good at this, as is the Nevada Bureau of Mines.

Acquiring and reading geological maps may be one of the bigger learning curves in becoming an intrepid rockhound. There are a few things to know about the world of maps that I wish I knew when I first started out:

1. There are many types of maps: topographic, hydrological, political, and so on. We are looking for *geological* maps. They're the ones that show the rock types underfoot. It's essential to use the term *geological* when searching for maps. On geological maps, the rock types are color coded—if it's wildly colorful, it's likely a geological map.

2. Maps come in all sizes. Yes, they come physically printed in many sizes, but what I mean is that maps are *drawn at many different scales*. Scale is one of the most important things to become comfortable with when looking at maps. There are geological maps that depict all of the United States and Canada, a single state, or an amorphous region within a state. In the western states, maps are often drawn as specific rectangular quadrants of land called *quadrangles*. Keep in mind that the Earth is round, so rectangular maps are mildly stretched to accommodate our human desire for edges and corners in a cornerless world. Geological quadrangle maps are my favorite because they provide a tremendous amount of detail for a small area.

3. Searching for the correct map can feel daunting. This is where picking up the phone or sending an email can really help. Both the USGS and state agencies have information officers who help the public locate the maps they need. Let's say I'm going to visit my sister in Reno, Nevada. I'll call the Nevada Bureau of Mines and ask for all the titles of geological maps that contain Reno and its outlying areas. I'll also check the USGS's National Map Database for geological maps of the Reno area. The USGS has done its best to compile maps from every state agency in one database, and sometimes it is much easier to find maps directly through them rather than state agencies.

4. Digital or print? Sometimes you don't get to decide. I've come across many maps that are only available on paper. Some are just old and have never been digitized. Others have been brought into the modern era. Digital versions of geological maps range from large, unwieldy PDFs that are slow to open on computers to wonderful .kmz files. These are zip-compressed .kml map files you can import into Google Earth and lay right on top of satellite imagery. Pretty awesome! I recommend learning to look at maps this way, especially if you want to get a 3D perspective of a rock formation. However, nothing can beat a large, printed paper map in terms of detail and portability. When the power goes out or a battery dies, paper maps still work.

Let's walk through how I like to use geological maps to locate rocks. In the Mojave Desert of Southern California, I own a small five-acre parcel of private land that sits within a massive ocean of public land administered by the Bureau of Land Management. You may remember

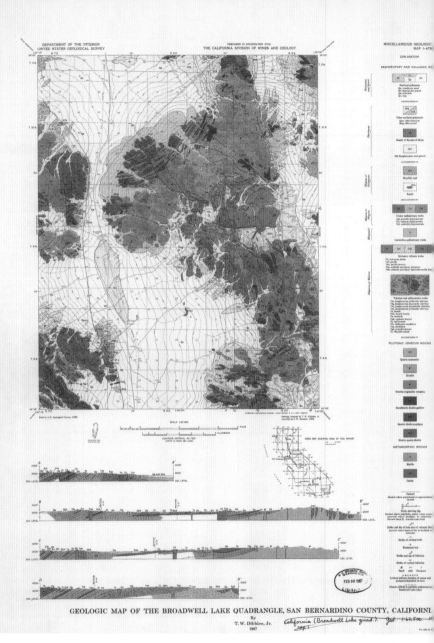

GEOLOGIC MAP OF THE BROADWELL LAKE QUADRANGLE, SAN BERNARDINO COUNTY, CALIFORNIA

By
T. W. Dibblee, Jr.
1967

me mentioning that I don't own the mineral rights to my land, so I can't collect rocks on it. However, I'm surrounded by millions of acres of public land where I'm free to rockhound. Because of the sheer size of the area, I knew I should find out what rocks were where before venturing far. Consulting a geological map of the area was my best bet. I looked to see if there were geological quadrangle maps available, which show a lot of detail. After some searching, I discovered that my parcel of land sits in the Broadwell Lake Map Quadrangle in Southern California, which, I noticed, was next door to the Cady Mountain Quadrangle—the Cady Mountains are a very famous rockhounding region, so I knew I wanted that map too. I searched the USGS National Map Database for geological maps of each quadrangle and found them. I looked at free PDF versions to make sure they were the right maps and then ordered paper copies of both.

These maps are large, and far too detailed to be legible in a book this size, but I still want you to see what they look like in their entirety. The large geological map pictured is of the Broadwell Lake Quadrangle (see page 128). My property sits at the very top boundary of the map in all that light yellow stuff, which is loose sediment that has eroded from the mountains nearby and washed into the broad valley below. On geological maps, rock types are color coded. Light yellow is always loose valley sediments. Each map color is also given an alphanumeric symbol so you can tie it to all the important information in the map key. (The map key is where all the juicy info is kept.) Remember my advice on becoming familiar with the geologic time scale? This is where that really comes in handy. If you acquire some sort of fun poster or mug with the geologic time scale on it, you'll notice it's wildly colorful. That's on purpose. All those colors have distinct scientific meaning. They're the colors the Geological Society of America has set as standards for mapping rock units. The colors have nothing to do with the colors of the rocks themselves. The colors are tied to time periods and material composition.

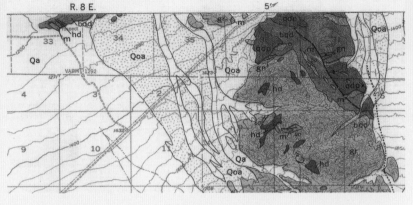

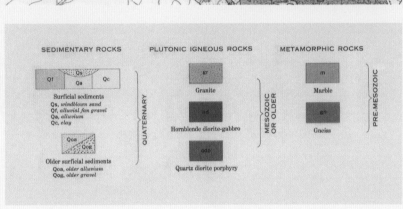

SEDIMENTARY ROCKS

Qf | Qs | Qc
| Qa |

Surficial sediments
Qs, *windblown sand*
Qf, *alluvial fan gravel*
Qa, *alluvium*
Qc, *clay*

Qoa / Qog

Older surficial sediments
Qoa, *older alluvium*
Qog, *older gravel*

QUATERNARY

PLUTONIC IGNEOUS ROCKS

gr
Granite

hd
Hornblende diorite-gabbro

qdp
Quartz diorite porphyry

MESOZOIC
OR OLDER

METAMORPHIC ROCKS

m
Marble

gn
Gneiss

PRE-MESOZOIC

The only problem with this clever system is that it's not friendly toward those who are color-blind. So the alphanumeric symbol tied to each color is paramount.

Let's use the map key on page 131 to find some rocks! Just across the valley from my parcel is a large block of pink and purple rock (gr, hd, qdp)—a mountain range rising from the pale-yellow sediments (Qa, Qoa). The map key tells me these are plutonic rocks that intruded into this part of the crust during the Mesozoic, the era of the dinosaurs. Perhaps it was the basement of a volcano system long gone? Scattered within this ancient pluton are wee bits of bright blue (m) and dark green (gn). The key tells me this stuff is even older than the pluton. It was there before the pluton showed up. The key also tells me these materials are metamorphic, meaning they were baked, but not melted. Perhaps the pluton is what baked these older rocks with all its intense molten heat? Either way, the metamorphic materials are what I'm interested in, particularly the bright blue stuff (m). The map key tells me it is marble— the result of limestone getting baked. I've recently gotten into sculpting with marble, so I've marked these bright blue outcrops on my map. Now that I know where they are, I'll see what jeep trails exist nearby and assess how much hiking I might have to do to get to them. Getting pieces large enough for carving means I won't want to carry them far. I can assess hiking and driving distances by consulting the scale bar on the map. Sometimes roads are marked on geological maps too. If not, I'll search recreation atlases and satellite imagery.

Of all the outcrops of marble on the map, I'll try to get to the most accessible one. There's one patch of blue right on the edge of the yellow—meaning the valley sediments slope right up to the marble deposit. I might be able to just drive right up to it!

If only paper maps were available to me, this is about as advanced as this process would get. I would note the general coordinates of the marble outcrops and study possible paths to get to them—and prepare myself for wayfinding in the desert. But to my delight, digital .kmz files were also available for the Broadwell Lake and Cady Mountains quadrangle maps. I was able to import the map files into Google Earth and lay them on top of satellite imagery and mark the exact coordinates of the marble. By making the geological map layer slightly transparent, I was able to get a strong sense of the topography since I could visualize the satellite imagery underneath. I could see roads and jeep trails, and I was able to measure the distance of the outcrop from my property and estimate mileages for some of the possible routes between them. Once out in the desert, I can plug those specific coordinates into a handheld GPS unit to help me stay on track. Sometimes this method almost feels like cheating, but it can be helpful when trying to locate a tiny deposit of rock hidden within a more massive formation, such as these wee pockets of marble floating in all that plutonic rock. When I go out to look for it, I'll still roll up my laminated paper map and bring it with me as backup.

✦✦✦✦✦

You've heard from me about my method of searching out specific types of rocks, but it's by no means the only way to go about it. I asked some of my favorite rockhounds to share their methods with you. Some folks are longtime devotees and others are still new to the hobby. Here's what they had to say:

"Rocks are everywhere. I've always told people to study geology if you don't want to ever be bored. If you're a geologist, or a serious rock-hound, even sand is interesting! I've always loved rockhounding in the desert, because the exposures are so good, but now that I live in the Pacific Northwest, I've become a fan of riverbanks and beaches. I have a pretty exten-sive library, and sometimes I use Mindat.org. But I find that when I go someplace that some-one else has suggested, I'm usually disap-pointed! My favorite thing to do is just head straight out into the desert, into some lonely mountain range, where maybe nobody has ever been. I may not find the "best" rocks, but the sense of discovery is more intense."

—**Martin Holden, Washington**

✢✢✢✢✢

"I'm a surface hunter through and through. I talk to friends and read books to get a gen-eral area and then look on Google Maps to find promising areas (dry creek beds, washes, and eroding hillsides mostly). There are only a few places I routinely go to, and I try to just explore over time and dial in where the loot is."

—**Cam Fortin, Idaho**

"My process for finding rocks is mainly searching beaches. Every beach has different stones, mostly local to the area, but you can also find some more interesting things that wash up after a big storm. I'm a lapidary artist. I look for color in my work. Patterns and textures are also important, so I tend to pick one of those things and stick to it on any given day."

—Marge Hinge, Rhode Island

✥ ✥ ✥ ✥ ✥

"I'm a lazy fossil hunter and rockhound. I do most of my work online before even going out. I check Facebook groups, online forums, social media, all to see what everyone else is doing. Of course, a geological map is absolutely essential. Since my home state of Texas is almost entirely privately owned, options are limited. It can make it difficult to find good spots, but easy to narrow down in a large state like mine. Unfortunately for me, I am drawn to igneous and metamorphic areas, which are not exactly abundant in Texas."

—Jessica Martin, Texas

"I look online, watch YouTube videos, read guidebooks, etc. Rockhoundresource.com is a great resource that has Google maps of rockhound sites for each state, and you can download it to your Google Maps app. Mindat.org is helpful for finding what minerals are in an area. I also use the Gaia GPS app for off-road navigation. (Don't use Google on dirt roads. It's inaccurate in remote places.) Gaia has different layers, one of which shows mine claims. Granted you can't dig at private claims, but it lets you know what is in that area. It also shows property lines and public land areas. Finally, and probably the best, I go into local rock shops and talk to the people who work there. They often know of places that are lesser known. I don't have any preference, but I find myself looking for rocks everywhere. You never know what you'll find. Just as long as I don't run into something while looking at the ground."

—Rachael Basye, Oregon

"I typically hear of an area or have seen an image of material someone else has found somewhere and then research that area online using Mindat.org, GIS, USGS, BLM, Google Earth, and many, many more. Before I leave, I'll use Google Earth to review the topology and search for outcrops I want to hunt along. I look for nearby camping facilities and check all the roads (dirt, gravel, and paved), then save it to offline maps for reference on my cell phone, even when there isn't connectivity. I tend to love the desert for how it helps my joints and pain…plus it's very beautiful in my opinion. I also truly love rockhounding rivers and streams on hot summer days for a multitude of reasons."

—Shaughn G. Belmore, Oregon

✤✤✤✤✤

"If I'm traveling anywhere, I typically do an internet search on what rocks may be around. If I can't find good info on exact locations or methods to find, I tend to go to the rivers and search through the gravel. I figure if it's around a river, some will end up there, and why not get wet while looking!"

—Kellan Smith, Washington

CHAPTER 4

Travel Smart

"Do your homework before you hit the road."

—Jay Stein, rockhound in Oregon

Travel is an inseparable part of the rockhounding hobby. Some trips can be a short jaunt to the brook that runs through the forest nearby, others a weeklong dirt-roading excursion through the sagebrush desert. No matter the intensity or length, all rockhounding forays need to be at least somewhat planned, if not exquisitely thought through. When it comes to getting outdoors, we all must start somewhere. But naivety about wayfinding, camping, dirt-road driving, and the elements can lead to all sorts of hijinks, or worse, real trouble! Plans change, the weather turns, roads disappear, tires flatten. These things will always happen, no matter how prepared or confident we are. What matters is that when we travel, we go out there as resourceful problem solvers and not total liabilities.

The most important part of trip planning starts at home. In the previous chapter, almost all the rockhounds you heard from mentioned doing research using books and the internet to plan trips. I highly recommend following their guidance. I do tons of research on the web before leaving for a trip. I estimate mileage, I look up rural gas stations (and call them to make sure they actually have gas), I note location coordinates, and I look at everything via satellite imagery on Google Earth. Digital resources are awesome at home, but they can be hard to take on the road. Yes, you can make a Google map and save an offline version to use when cell service disappears. Yes, you can use a GPS unit to guide you to coordinates in the middle of nowhere. Yes, if you still have service, you can ask your phone for directions. But what happens if your phone dies? Or falls in the river? Or you leave it on the roof and drive off into the sunset? What happens if you forget batteries for your handheld GPS? Or the puppy chews the charger? What happens to your trip if all that info you need to get to that cool place is now completely unavailable? And you have no idea where you are?

If you had an awesome, up-to-date paper atlas, and you had written everything down on that paper atlas before you left home, you'd be just fine. This atlas is different from your geological map. This atlas details all the roads and towns and topography in the area you're traveling through. When it comes to paper atlases, there is no better way to get around. They don't run out of batteries. You can drop them, sit on them, and write notes on them. (Just keep them away from the wind.) The learning curve with paper atlases is quick. The answers are always in the map key. All you need to do is pay attention to where you are on the map as you use it to travel. Even if you have a giant geological map with you, that thing is too huge to navigate with. Geological maps aren't very good about marking roads or landmarks, and yours might have been mapped back in 1960. (That's fine because the rocks are probably still the same, but the roads are not!) What you want in addition to a geological map is a trusty, up-to-date road atlas, specifically a *recreation atlas*. For federal public lands, the Bureau of Land Management releases recreation atlases of its various regions. Some US states and Canadian provinces publish them too. Good recreation atlases feature public land boundaries, paved and dirt roads, campgrounds, boat ramps, points of interest, and access to services like food, fuel, and hospitals. For the western states, I'm particularly fond of the recreation atlases published by Benchmark Maps. I especially love their large-format gazetteers, which have extremely detailed road maps that include the surface quality of dirt roads. I can tell the difference between a dirt road I can take a Prius down and dirt road that will be challenging even in a lifted 4x4. Federal, state, and private land are color coded as well. This is extraordinarily helpful when I'm out rockhounding. It keeps me off private land.

My process for using paper maps and atlases goes like this: I do a bunch of research at home on the web, in books, on geological maps, and in my atlases. Then I write everything down in my Benchmark atlas: I put stars on relative locations of rock sources, mark places I might want to camp, and highlight alternative routes with a highlighter. I put question marks on things that are mysterious and write notes in the column and dog-ear pages. I keep my geological maps rolled up with me for reference, but my paper recreation atlas is my compendium of information. I also bring my phone and GPS with me, but I don't count on them as the only tools to help me get around.

A strong word of caution: do not ever ask Google for directions in remote and wild areas, especially in the desert. Google and all other basic navigation apps will con-fuse drainages and washes for dirt roads, cliffs for non-cliffs, private driveways for public roads, and so on. Please, if you only take one thing away from this book (OK, two things, since this is my second ask), let it be this. Besides, it's so wonderfully liberating to practice real-world navigation with paper maps while learning to read the landscape as you drive. That odometer in your car is your new best friend. Almost all vehicles have the function to set and reset odometers to track mileage as you drive. Mark your miles and mark your turns. If you know exactly how you got in somewhere, you know exactly how to get out. If I find myself in an area with a web of roads, like the maze of oil roads in the Uintah Basin, I get out my notebook and mark the mileage and direction of each turn using my odometer and vehicle compass. I make a "turn map" in a spare notebook as I go along. Alternatively, I keep highlighters in my glove box so I can highlight the route directly onto the atlas as we drive.

Furthermore, if I'm exploring a large, expansive area (like badlands), I try to park my vehicle on some sort of rise, not in a dip, so I don't lose sight of it. In forests, especially level ones, it can be so easy to lose your way. GPS units can really come in handy in places like this. Mark your coordinates at the car and use your GPS to find your way back to them. Alternatively, when in forests, I always follow river and creek beds so that I don't get lost. That's where all the easy rocks are, anyway.

The whole purpose of preparedness is to have fun and avoid stressful situations. We all know the basics of this: have plenty of water, food, fuel, and ways to stay warm. Keep a well-stocked first aid kit and your

prescriptions with you. Never plan on hav-
ing reliable cell service. Bring a gas can.
Tell someone where you've gone to and
when you'll be back. We all know that plans
change. Mother Earth throws curveballs.
And she throws them often! Our responsibility traveling into remote regions goes
beyond good planning. We also need to
be able to perform a reasonable amount of

self-rescue if things go wrong. When the weather changes, so must the plan. If the road becomes too treacherous, get out and walk. If the car starts acting up, turn it around. Flat tires are the most common rockhounding *womp-womp*, but they're easy to fix. There are some things I like to keep in my vehicle to help me out of the most basic trouble. Let me share them with you:

✣ I make sure my tire-changing equipment is all there before leaving. I also have a breaker bar—which works astronomically better than a tire iron—to get lug nuts off. Breaker bars are cheap at the auto parts store. Ask one of the employees to show you how they work. I also keep common lug nut–sized sockets in my road kit to help out other kinds of cars too. If you want to travel on rough roads, you need to feel comfortable changing flats. If you've never done it, find an experienced friend and practice in your driveway! Really!

- I carry some 1 × 2 ft. plywood scraps to place under my car jack if we're changing a tire on soft sediment. It prevents the jack from sinking or collapsing into the dirt and dropping the car onto me.

- I keep traction mats to put under the tires if I get stuck in mud or deep sand. Other people keep rolls of scrap carpeting, strips of plywood, or even the floor carpets from the vehicle's driver and passenger seats. I always have a small folding shovel or two.

- It's not as necessary as the other items, but I keep a portable air compressor in my vehicle. It connects to the engine battery for power. I like to drop my tire pressure on long drives through deep sand or rough terrain to help prevent sinking and flat tires. When I get back out to pavement, I use the portable compressor to fill the tires back up. This little compressor has also saved the day numerous times for folks I find stranded on the road with a tire losing air.

- My stepdad gifted me a cool portable jump starter in case my car battery drains or dies. I charge it up before each trip. I can charge phones and other devices off it as well. It's not necessary to carry one of these but it feels comforting. The better option is to test your car battery before a trip to a remote area. Swing by the auto shop before a big trip and have the folks there juice it up for you.

- I keep a nice first aid kit *in an easily reachable place.* In my last truck, I kept a kit tucked away in a compartment under a jump seat. Many years back, I was collecting obsidian by myself at the Mono Craters in California and got the gash of a lifetime across my right hand. Obsidian is glass, after all, and I wasn't wearing gloves! I ran to the truck, only to realize that all my camping equipment and a heavy cooler were tightly stacked on top of the jump seat. I had to unpack the whole truck cabin left-handed just to get to the kit. Oy vey!

Sometimes things happen beyond self-rescue. A starter or alternator dies, a radiator becomes a geyser, the old engine seizes. Because of this, you should never drive into a remote area if you're not willing to walk back out. This is a lesson I never want you to learn the hard way.

For me, hearing stories of other people's fiascos really helps prepare for scenarios that would never even cross my mind— like making sure your passenger would be able to drive your vehicle if you got hurt. This happened to my friend Charlie on an obsidian-collecting trip (a trend?). It was a gusty day, he got rock dust in his eyes, and couldn't see. Pretty bad, I guess. And his wife didn't know how to drive a manual transmission truck. So, blind, he had to teach her how to drive it down a rocky slope. They slowly lurched their way down the mountain and eventually made it home. Talk about a driving lesson! I asked some of my other rockhounding friends to share some stories of lessons they learned the hard way. As stories sometimes stick better than plain old advice, it's my hope that you can learn from their mishaps and have a safer trip the next time you head out to search for rocks!

✦✦✦✦✦

"My partner, Mo, and I went on our first rock-hunting trip down in southeastern Oregon. Our first stop was Plush. We wanted to find sunstones. While driving more or less aimlessly out in remote, uninhabited desert, on roads that would test even the most intrepid traveler, you guessed it, we got a flat. We were in the middle of nowhere. I started to jack up our Mitsubishi Montero to change the tire when the car shifted, and the jack slipped in the sand and got lodged under the vehicle's frame. There was no way that we could free up the jack at this point. We realized, OK, we're screwed. After about two hours we decided to try and hike out. We walked about a quarter mile when this old-timer came up in a '54 Chevy truck with a lift on the bed. He drove us to our car, pulled the car out of the sand, changed the tire, and got back into his truck. Before he drove off, he stuck his head out the window and said, 'You folks need to be careful out here, ya hear me?' Then he was gone and we were saved."

—**Tom Provenzano, Oregon**

"I drive a Ford Mustang. That should be enough to tell you where this story is going. One day I decided to head to a favorite river in rural Texas where I (and others) often hunt for fossils. Since you have to enter the river near a bridge, and the area is all farmland and pasture, there's no parking lot or paved shoulders. I parked off the road as I always had before, but to get far enough off the road to avoid being hit by any incoming cars or tractors, I overcompensated and parked a bit too far off the road. When I was ready to leave, I found I was stuck, and my efforts to get out only put me further into the ditch. Thanks to AAA and many friendly farmers offering a hand, I was finally able to get out in just an hour or two!"

—Jessica Martin, Texas

✢✢✢✢✢

"I remember one time we tried to find a public dig site using GPS coordinates and it was a wild goose chase! The navigation tried to take us off the road and then over a steep ravine! So don't always trust the navigation directions."

—Laurie Engelhardt, Washington

"I was driving on unmarked roads in BLM badlands and I took a wrong turn. I ended up on a road so rough that even though my vehicle 'made it,' I was dragging a shock when I got to the dig site. I had to drive on washboard roads and all the way home with no suspension."

—Lucille Cashin, Oregon

✢✢✢✢✢

"Decades ago, I found some blotchy copper minerals in a mine tailings pile in the Potosi Range outside of Las Vegas. Once home, I cut a piece and it had the most beautiful blue azurite spheres suspended in green malachite. I've combed every dirt road in that area over the last 30 years and haven't found the location again. Always make maps and take GPS recordings of your finds."

—James Palmer, Oregon

"Once when I was a kid, I was rockhounding near Jacumba, California, on the Mexican border. I was on top of a small mountain called Boundary Peak when a thunderstorm blew up. As usual, I had my Estwing rock hammer in my hand (I used to walk along twirling it, like a gunslinger twirling his six shooter). Suddenly, a bolt of lightning crashed into the rocks about ten yards away. My hammer started glowing green and felt warm in my hand despite the rubber grip. Instinctively, I threw the hammer as far away from myself as I possibly could and started running down the mountain. I loved rocks, and I loved my rock hammer. But not enough to die for them!"

—Martin Holden, Washington

✣✣✣✣✣

"About ten years ago I went with my dad to find zircon crystals near Zirconia, North Carolina. The zircon mine had closed almost a hundred years ago so we didn't know where the tailings were located exactly, but we had a general idea. Very, very rural Appalachia. Twisty mountain roads. We pulled up to a property with several people sitting on the

porch, and we asked about the zircon mine. A young man slowly held up his shotgun and told us we were in the wrong place. "There ain't no mines around here," I remember him saying. We thanked him and retreated to our car. Later that day, I was damn near eaten by a vicious, wild dog as I walked along a road-side searching for signs of mine tailings…

We didn't find any zircon that day. But the next year, my dad had some better information from people in his church, and we were admitted onto a property by a man who recalled digging zircons out of the stream gravel as a kid. We went to the creek, spotted some zircons in the creek bed, and collected a few buckets of gravel. We sieved it at home and found lots of large, beautiful zircon crystals. Did it again the next year too! Moral of the story: 'general' ideas as to the location of an old mine is not at all sufficient, and the scale of the land can be confusing even for seasoned mappers. Talk to the right people to narrow your search. And finally, don't expect people to enjoy the fact that you are wandering around their home looking for treasure."

—**Barry Walker Jr., Oregon**

CHAPTER 5

Tools
of
the Trade

"There's nothing like finding a cool rock,
bringing it home, and cutting it open. This
rock sat covered by earth for millions of
years and now you are the very first person to
see inside of it. The experience is something
that moves you in ways nothing else in
nature can muster."

—Tom Provenzano, rockhound in Oregon

Rockhounding is probably most synonymous with the emblem of the rock pick crossed with the shovel. It's an old symbology borrowed from geology and mining representing the tools of the trade. Rocks are hard. It takes a lot of power to break them up. The right tools can make all the difference. In this chapter we'll take a flyover of the tools rockhounds use to extract rock from the ground and make it look pretty.

When it comes to extraction, I very rarely use tools. I'm a surface collector. That means I like to walk around and see what's on the ground. I dislike digging. It feels disruptive and damaging to the soil. My preference is to collect rocks gently from the surface. Instead of a rock hammer and shovel, my two most beloved collecting tools are an old canvas fly-fishing vest with slouchy pockets I can fill with stones and an old canvas bucket. The canvas buckets collapse when empty and are wonderfully light—much preferable to hiking around with a hard plastic bucket. Buckets are kind of like purses. The bigger they are, the more you stuff them with things you don't need. I try to be extremely selective when I'm out surface collecting to keep the load light.

Most rockhounds, however, live for the dig. To them, nothing is better than a day in a trench with a good shovel or along a rock face with a chisel and hammer. There's a variety of hand tools that help break up rock. It's good to know why they all look a little different from one another and what purpose they're best used for. I'm going to trick you into thinking this section is about fun tools, but secretly it's about tool safety. When you hit something hard, like a rock, with something almost as hard, like a hammer, the rock cracks and shatters. Small shards of jagged, sharp material fly everywhere. Swinging heavy things at rocks without wearing good safety glasses is phenomenally stupid. Furthermore, when you hold a chisel in your hand and hit it with something hard, like a hammer, I hope you will always consider where your hand is. I hope you will be wearing thick gloves, and that you will swing your hammer with intrepid awareness.

The most important hand tool is not the rock hammer, as one might assume. It's actually the **claw** and the **shovel**. Not big ones, but rather small ones you don't mind hiking with. These will help you flip stubborn rock specimens out of the ground. Perhaps even more excellent is a specialized **rock pick** that looks less like a hammer and more like a blunt garden spade (see above). It's designed for clawing away dirt on one side and chipping away at hard-packed dirt with more precision on the other side. This tool is not made for hitting rocks, it's made for moving earth out of the way, which it does it with great efficiency.

When it comes to breaking rock apart—or hard-rock mining—specialized hand tools are required. You're going to want to purchase tools manufactured from hardened steel. One of the best companies to buy from is Estwing. They make handsome rock hammers and chisels that are available at most hardware stores. There's a variety of hand tools that all serve specific functions. A **rock hammer**, or **geology pick**, looks the most similar to a nail hammer. These are made to knock portions of rock off a larger outcrop—but not with precision. They're mostly used by geologists to collect hand specimens for observation. Blows from a rock hammer tend to shatter rock indiscriminately, but that's what geologists are hoping to do. So if you're trying to carefully break a special piece of stone out of a rocky cliff, a rock hammer/geology pick is *not* the best tool.

For more precision, use two hand tools: a **chisel** and a **drilling hammer**. Drilling hammers look like small sledgehammers.

A two-pound drilling hammer is the ideal weight for hard-rock mining. When shopping for chisels you'll notice a variety of tips. Pointed chisels are far superior to flat chisels for breaking hard rock. They focus the power of your hit into a single point. Flat chisels spread that power out and are better for splitting soft stones like limestone and slate along a plane. Think of the shape of the chisel tip as an illustration of how the blow is delivered to the rock surface. Make sure your chisels are manufactured with hardened steel for breaking rock and are harmonious in size with your hammer weight. (Someone at the hardware store can help you with this.) When swinging a hammer at a chisel, it's important to have control, but also a relaxed grip. If you're squeezing the hammer, your hand will absorb most of the impact, which can lead to strain and injuries. Loosen your grip a little and remain focused on your swing. If you're going to be working with extremely hard rock, like agate or jasper, you may want to consider investing in carbide-tipped chisels. Trow & Holden

Company in Vermont makes some of the finest masonry and rock chisels on Earth (with a price tag to match!). No matter what your chisels are made from, you will have to sharpen them occasionally. A machine shop can help you with this if you don't have a bench grinder at home. Another important hard-rock tool is a **pry bar**. Pry bars can be a real lifesaver when trying to work large sections of rock apart. Use them like a wedge. If you're splitting something so big that you need a pry bar, I hope you won't be digging alone! Here's the most important thing to know about hard-rock mining: if you're working to get a specific material out of hard rock, like a pocket of crystals, an agate nodule, or a vein of gold in quartz, you work *around* that material. Go wide! The moment you strike the special material directly with your chisel and hammer, it will shatter, and your hard work will have been in vain. Hard-rock mining is slow, patient, and strenuous work!

Daintier tools are needed for breaking up softer rocks, especially splitting sedimentary rock layers for fossils. For splitting

large mudstone or limestone slabs, I like to use a **smaller hammer** and a **paint scraper** with a blunt end on the handle. I tap around the slab's circumference until I gently split layers apart. This helps reveal delicate fossils like leaves or insects trapped in the fine layers. Some rock shops also sell special fossil-splitting wedges, which work wonderfully, but aren't easy to come by (see above). For small slabs of stone, just delicately working the sedimentary layers apart with a pocketknife can do the trick. It's also nice to have a jeweler's loupe or **hand lens**. These are small magnifiers that help you see fine details in rocks and fossils. I used to think they were dorky but now I don't go anywhere without one. As I've become more adept at identifying rocks, my hand lens has become indispensable.

If you're someone who's interested in prospecting for gold, the variety of necessary tools grows. There's the simple art of panning for gold nuggets in rivers that wash out of plutonic mountains, which is quite meditative and inexpensive. **Gold pans** are sieves used to settle out heavy metal nuggets from less dense gravels. You can graduate to **sluice boxes** (see opposite page), but this requires setting up a dredge-mining operation on the river, which is ecologically disruptive (and requires permits). Perhaps you're more of a desert rat, and you want to search the rocky outcrops for veins of ore looked over in the Gold Rush. This is where the credit card really comes out. These days, serious

prospectors search for ores with an **XRF analyzer gun.** It's literally a futuristic ray gun that you point at the ground and it tells you the elemental composition of the rocks, especially values for metals like gold, silver, copper, molybdenum, nickel, tin, and so on—a hundred percent cheating if you ask me! But, golly, do they work! If you get some good numbers back, it's time to stake a claim and buy yourself a backhoe. Kiss the beautiful sagebrush goodbye and yeehaw yourself into the sunset.

Lapidary Practice

Getting rocks out of the ground can be a lot of work. The great motivator for all of it is getting to take the rocks home and making something beautiful out of them. Let me tell you about my setup for processing rocks once I get them home. If the rock is for sculpture, it goes out on a workbench where I will cut away at it with a diamond-bladed angle grinder, a pneumatic air hammer, and hand tools like small hammers, chisels, and files. The process is mostly *dry*, which means there is no water involved and it creates a lot of dust.

If I want to make jewelry (or something small, like this box made from offcuts and scraps from jewelry I have made), I'll use lapidary equipment. The craft of cutting small stones (lapidary) goes back thousands of years. Similar to stone sculpture, the technologies used to cut rocks haven't advanced much over the eons. It's mainly just breaking rocks up and grinding them down. Lapidary work is mostly a wet craft. Large equipment is lubricated with oil and smaller equipment is lubricated with water. It's messy! The first step is to get the rock cut down into thin slabs. If the rock is large, it goes into a slab saw first (see page 172, top). These saws take a big rock and cut slices from it, just like a deli slicer cuts ham. Once I have the slabs cut, I make finer cuts out of each slab on a trim saw. Trim saws are much smaller rock saws that work very similarly to a tile saw. If I want my final product to be flat, I will then work my piece of stone on a machine called a flat lap, which has interchangeable flat grinding discs from coarse to fine. Or, if I want to make a rounded, domed-shaped stone for jewelry (think of the smooth round stones commonly set in rings), I will work it down on a cabochon machine. "Cabbing machines," as they're more often called, are the most popular lapidary tool and a great investment. They're easy to use and tend to keep their value over time. Cabochon machines are user friendly enough that if you bought one you could teach yourself to use it in just a day. If you want to get started in the lapidary craft, a six- or eight-inch trim saw and a six- or eight-inch cabbing machine are all you need!

Faceting

What about tiny gemstones? Perhaps you found some beautiful crystals while out rockhounding, or a lovely carnelian agate. You may want to consider faceting them! This is the craft of geometrically cutting a small crystal or stone to make it sparkle. Faceting is done on fantastically engineered desktop machines similar to a flat lap but fancier. There's much more of a learning curve to begin faceting than say, making a cabochon. The best way to learn faceting is to find a mentor who can teach you. Faceting is all about following recipes. Most gem cutters follow standard designs that are based on radial symmetry. It's a lot like a baking recipe, with strict measurements, but instead of cups of flour, it's degrees within a circle and angles within a plane. Purchasing a faceting machine

requires a little bit more of an investment, but it pays off if you sell the gemstones you cut.

If you just want to do "specimen work," meaning you'd just like to polish part of a rock you found, you can do it with some simple tools. Stone-specific grinding discs can be fitted to angle grinders and stone-specific sandpaper can be fitted on a belt sander. I even know one artist who polishes his specimens by hand with carbide sandpaper. The most important thing to know is that the dust created by grinding and sanding rocks is *extremely dangerous*. Grinding and sanding tasks should be done either outdoors or in a space engineered for filtering rock dust. No matter how good your dust collection and ventilation systems are, you still need to wear eye protection and a functioning respirator rated N95 or higher. No excuses! Inhaling silica dust from rocks can lead to a dreadful disease called silicosis later in life. Even doing small lapidary work on trim saws and cabbing machines requires wearing a mask. Rock particles in the air, wet or dry, are no joke when they get into the lungs.

There are other concerns beyond silica. Chatoyant rocks contain asbestos. Chatoyance is a beautiful, fibrous, optical reflectance popular with rockhounds and gemstone lovers. Unfortunately, the optical phenomena is caused by asbestos fibers.

Asbestos rocks come to market with common names like tiger's eye, tiger iron, cat's eye, and pietersite. Inhaling small amounts of asbestos can lead to mesothelioma later in life. I won't even let asbestos-containing rocks into my studio. No way! The few times people have gifted me tiger's eye, I brought it to hazardous waste at the dump. Dust from copper stones like malachite and turquoise can cause real problems in the lungs as well. Diseases from rock dusts don't manifest right away. They form as your lungs slowly reject stone particles, forming scar tissue and cancerous growths around them until they interfere with pulmonary function. Because of these hazards, I can't stress the importance of safety gear enough. This information may feel intimidating, however, if you practice good personal safety while cutting rocks, you can pursue the hobby for a lifetime without threat of respiratory illness. One of my favorite resources to consult for material safety in the studio is *Health Hazards Manual for Artists* by Michael McCann, PhD. If you are a studio artist or hobbyist, go get yourself a copy!

CHAPTER 6

Build a Meaningful Collection

"I treasure my rock collection because it
represents the adventures I've taken."

—Lucille Cashin, rockhound in Oregon

A few years ago my friend Joel posted a picture of a man sitting in a shack with his rock collection (see opposite page). The man was surrounded by thousands of stones that all seemed to be the same. Roundish, unremarkable cobbles, probably mudstones. Plain, old rocks. But on closer inspection, you could see that there was something special about them. They weren't unremarkable at all. They were exquisitely categorized by character and shape. Round shapes, big kidney shapes, little kidney shapes, flat shapes, and so on. All organized meticulously and lovingly. Never had I seen such ordinary rocks presented in such an extraordinary way! It really warped my idea of what a rock collection could be—in the best way. I started thinking more deeply about how many different approaches there can be in rock collecting. How personal it can be. Our collections are an extension of ourselves. Of what we consider beautiful or interesting. And in the case of this man, Luigi Lineri of Italy, it's a heartwarming reflection of a lifelong pursuit. The photo of Luigi, taken by photographer Colin Dutton, lit the flame of a very big candle for me. There's so much darkness in the human world. We put so much energy into chaos and harm. Seeing Luigi's brand of devotion reminded me that rock collecting can be such a meaningful way to spend one's time on Earth, appreciating the very fabric our planet is woven from.

A rock collection doesn't have to be massive, like Luigi's, to be impressive. Some of the coolest collections I've seen fill only a bucket or two. Take *rock feasts*, for example. These are collections of rocks that look like food, laid out banquet style, candelabras and all (see opposite page). Rock feasts are usually displayed at rock and gem shows organized by local rock clubs. People who collect rocks that look like food spend years searching for the perfect "fried egg" or "flank steak" to add to the table. Imagine their delight when wandering along a riverbed and they find a big cobble that looks just like a loaf of bread! Such joy!

Some rock collections can even be truly miniscule. My friend Nicky Kriara collects

sand. She has folks send her sand samples from all over so she can photograph them under a microscope. It's likely her whole collection fits in a shoebox. My friend Andrew Howard likes to collect strangely shaped mudstone concretions. He keeps just a few, and the shapes are just wild! It's hard to believe they're rocks at all! Kirsten Southwell, one of my closest rockhounding comrades, produces high-resolution scans of her modest rock collection, pictured opposite, and uses the imagery on printed fabrics for apparel. My personal rock collection is also small—if you can believe that. If I gathered my whole collection together it would fill two five-gallon buckets. I have a few large rocks and petrified logs that I have collected for the garden, but the rest are specimens that fit in the palm of my hand. Almost all of them are "ugly rocks." What I mean is, they're not that colorful, or crystal-encrusted, or rare. The great majority of them mark a special place or an interesting piece of geological history. I write notes and coordinates on many of them with a sharpie. My collection is a catalog of memories and ideas. A few are rocks I've been gifted or inherited. My uncle P. J. sent me some magnificent feldspar crystals from Mount Erebus in Antarctica. He lived at South Pole Station doing physics for many, many years. When my grandfather passed, my mother shipped me a small bag of his stones, all small and ordinary. To my surprise, my grandfather had written a date or location on each stone with a marker, just like I do! After some research, I discovered the dates were anniversaries, and the locations were beaches that my grandparents loved to go to. Each stone was a memento from their romance over the decades. My grandmother died relatively young from cancer, and my grandfather openly grieved for her for the remainder of his life. This small bag of rocks I inherited after his passing is probably the most special in my whole collection.

The rocks we gather over our lifetime are expressions, devotions, archives, remembrances, investigations, curiosities, and obsessions. Collecting can be a deeply personal hobby. And it can be a delightfully social hobby too. If you live in a big town or city, one of the best ways to engage with the rockhounding community is to join your local rock club. Sometimes they're called gem and mineral societies, but they all fall under one giant umbrella of connected groups. To find out if there's a club near you, check in with the American Federation of Mineralogical Societies in the United States (amfed.org) and similar federations in Canada. The mission of rock clubs and societies is to bring new collectors into the fold, set up field trips and lapidary lessons, and encourage good stewardship and collecting ethics. Joining my local rock club transformed my life! I jumped in as a mildly curious person, and within a few months had quit my job to pursue rocks full time. The club members taught me how to cut rocks, took me on field trips to find beautiful material, and welcomed me into a world I hadn't known existed. I can't promise that joining a rock club will lead you on an illuminated path of dynamic personal transformation, but I can promise you that you will make new friends and have tons of fun.

There's so much to love about the rock-hounding community, but we also have our battles. A true thing: not everyone in the world of rockhounding takes it easy on Mother Earth. In fact, rockhounds have done a lot of damage over the decades, especially in places where beautiful material used to be abundant. I've been to a lot of yard sales held after an old rockhound passes away. It's the same scene every time—a garage filled to the ceiling with rocks. Fifty-five-gallon drums scattered in the yard, overflowing with rocks. Boxes, bags, buckets, trays, shelves, dog bowls, cookie jars overflowing with rocks. But none meticulously cared for like the collection of Luigi Lineri. These collections are hoards. And out in the wild, there are messes of open pits and empty trenches to match them. I went to one such sale a few years back where I witnessed the worst-case scenario. An old rockhound had hoarded *many tons* of Succor Creek thundereggs, a famous Oregon rock. It's likely he had dug them all up decades ago.

Just the week before, I had been digging at Succor Creek with my rock club. We found almost nothing. It was so disappointing! Apparently all the thundereggs were sitting in this old timer's garage, collecting dust! It was heartbreaking and frustrating. I wanted to buy the whole lot and bury it back in the ground at Succor Creek.

Experiences like these unsettle me in the same way seeing collections like Luigi Lineri's inspire me. The extreme swing between these two emotions is what made me want to write this book. I wanted to shout to the world that this is such a wonderful hobby, but it also needs to change. I love that every human I know has at least a few special rocks in their home. Collecting is a deeply instinctual behavior. Think about how babies grab pebbles and eat fistfuls of dirt. They can't help it! We grow older and those habits don't really change, do they? Picking up a rock is hardwired. Rocks are instinctually treasure to us. That's why it's so important that we build a meaningful collection, one that represents memories, ideas, or curiosities. And that we collect rocks with intention. There's nothing worse than letting our finds languish in a bucket in the garage. It's a cruel fate for such storied and ancient material!

That's something I love about rock clubs. They work hard to bring a better face to the hobby. The good people of the Mt. Hood Rock Club in Gresham, Oregon, are the ones that brought me into the fold. There's a lot of wisdom in that group. As I was writing this book, I asked them to talk to me about what they want to see change about the hobby. If there were any advice they would want to give to new rockhounds, what would it be? Stephen Petkovsek, our former club president, said, "The Rockhound Code of Ethics is a great place to start. Don't be a hoarder. Only take what you can reasonably use. Share your knowledge with others. Help when asked." My friend Kellan Smith said, "Respect the earth. Leave the area as it was, or nicer, and leave some rocks for others to appreciate." Fellow club member Shaughn Belmore would tell you to "be selective about what you take home. Only collect material that you know you will do something with. If you cannot picture what you will do with that stone, then leave it behind." Doing something with a rock can mean anything. You can display it, polish it, sell it, gift it, or stash it in a treasure box. It just needs to be meaningful to you. Every rock is a story, after all, and diving into the hobby of rockhounding tosses us into the infinite pond of geological time. Rocks offer us a glimpse into an Earth very different from the one we see around us today. They reveal the drama and cataclysms of the past, sometimes making our own human problems seem inconsequential. "When I despair the state of things, rocks always offer some comfort," writes author and geologist Marcia Bjornerud in her wonderful book *Reading the Rocks: The Autobiography of the Earth*. The hobby of rockhounding offers us a wonderful reset from the troubles of *the now*, but it doesn't let us off the hook completely. As collectors, we need to tread lightly so that we can promise the generations ahead of us that beautiful rocks will still be everywhere.

Rock Collecting Resources

US COLLECTING LAW RESOURCES:

✧ ✧ ✧ ✧ ✧

BUREAU OF LAND MANAGEMENT

"Can I Keep This?," http://www.blm.gov/Learn/
Can-I-Keep-This.

"FAQs on Meteorites on Public Land,"
https://www.blm.gov/sites/blm.gov/files/
uploads/IM2012-182_att1.pdf.

"Monuments, Conservation Areas and Similar
Designations," http://www.blm.gov/programs/
national-conservation-lands/monuments-ncas.

"Special Planning Designations," http://www.blm.
gov/programs/planning-and-nepa/planning-101/
special-planning-designations.

"Mineral & Land Records System (MLRS),"
http://www.blm.gov/services/land-records/mlrs.

USDA FOREST SERVICE

"Rockhounding Guide," https://www.fs.usda
.gov/sites/default/files/fs_media/fs_document/
rhbrochureFS1091.pdf

"Unpatented Mining Claims," https://www.fs.usda.
gov/Internet/FSE_DOCUMENTS/fseprd1058739.pdf.

NATIONAL PARK SERVICE

"Permits," http://www.nps.gov/subjects/geology/
permits.htm.

SOCIETY FOR AMERICAN ARCHAEOLOGY

"Archaeology Law and Ethics," http://www.saa.org/about-archaeology/archaeology-law-ethics.

DOUG'S ARCHAEOLOGY

"Is It Legal to Collect Arrowheads on Federal Land?," July 30, 2019, http://dougsarchaeology.wordpress.com/2013/06/03/is-it-legal-to-collect-arrowheads-on-federal-land.

CANADA COLLECTING LAW RESOURCES:

✤✤✤✤✤

"The Minerals and Metals Policy of the Government of Canada," Natural Resources Canada, http://natural-resources.canada.ca/science-data/science-research/earth-sciences/earth-sciences-resources/earth-sciences-federal-programs/minerals-and-metals-policy-government-canada/8690.

"Rockhounding North America," Calgary Rock and Lapidary Club, https://crlc.ca/wp-content/uploads/2016/05/fieldtripsweb.pdf.

ROCKHOUNDING & GEOLOGY REFERENCE BOOKS:

✤✤✤✤✤

Bates, Robert L., and Julia A. Jackson. *Dictionary of Geological Terms*. Garden City, NY: Anchor Press/Doubleday, 1984.

Holden, Martin. *The Encyclopedia of Gemstones and Minerals*. New York: Facts On File, 1991.

Falcon Guides Rockhounding Series. Guilford, Connecticut: Falcon Guides.

Gem Trails Guides. Upland, CA: Gem Guides Book Company.

Luedtke, Barbara E. *An Archaeologist's Guide to Chert and Flint*. Los Angeles: UCLA Cotsen Institute of Archaeology Press, 1992.

McCann, Michael, and Angela Babin. *Health Hazards Manual for Artists*. Guilford, CT: Lyons Press, 2008.

Roadside Geology Series. Missoula, Montana: Mountain Press Publishing Company.

LITERARY PERSPECTIVES ON ROCKS:

✢✢✢✢✢

Bjornerud, Marcia. *Reading the Rocks: The Autobiography of the Earth*. New York: Basicbooks, 2006.

Childs, Craig. *Finders Keepers: A Tale of Archaeological Plunder and Obsession*. New York: Back Bay Books, 2013.

Macfarlane, Robert. *Underland: A Deep Time Journey*. New York: W. W. Norton & Company, 2019.

McPhee, John. *Annals of the Former World*. New York: Farrar, Straus & Giroux, 1998.

THE ROCK.
OF FAITH.
AND TRUTH.
NATURE.IS.GOD.
THE.KEY.TO.LIFE.
IS. CONTACT.
EVOLUTION.IS.THE.MOTHER
AND FATHER OF MANKIND.
WITHOUT THEM.WE.BE.NOTHING.

JOHN. SAMUELSON.
1927.

Acknowledgments

The creation of this book is entirely thanks to Amelia Rina, who believed in my vision, and to Holly La Due, my editor at Princeton Architectural Press, who helped bring that vision to life. Thanks also go to wonderful copy editors Kristen Hewitt and John Son, and talented designer Natalie Snodgrass. The knowledge compiled within these pages is mine but gained only through the guidance and comradery of the fine people of the Mt. Hood Rock Club in Gresham, Oregon, and the Butte Gem & Mineral Club in Butte, Montana. I consider myself humbled to be a member of both venerable clubs. Thanks always go to my beloved wife, Lisa Ward, who fills my life with beauty, and to my devoted friend Kirsten Southwell for her guidance, encouragement, and enthusiasm. I send an echoing thank you to the pantheon of authors who have laid the groundwork for rockhounding guidebooks over the decades. I collect them all.

Credits

All images not listed courtesy
Alison Jean Cole

Cover images:
Alison Jean Cole

2: Arthur Rothstein / Library
of Congress
9: James St John
37, right: Printmaker unknown
/ Library of Congress
40: Arthur Rothstein / Library
of Congress
42: Photographer unknown /
Library of Congress
45: Kent G. Budge / Wikimedia
Commons
60: Photographer unknown /
National Park Service
82: Orra White Hitchcock /
Archives & Special Collections
at Amherst College
89, top: United States
Geological Survey / Wikimedia
Commons
89, bottom: James St. John /
Flickr
92, top and bottom: James
St. John / Flickr
95: United States Geological
Survey / Wikimedia Commons
96: Pietro Fabris / Wellcome
Collection
98, top: Photographer
unknown / National Archives
98, bottom left: Photographer
unknown / National Park
Service

98, bottom right: James
St. John / Flickr
104: James St. John / Flickr
106, top: James St. John /
Flickr
111: United States Geological
Survey / Wikimedia Commons
115: Nast & Martin / Library
of Congress
118: United States Census
Office and Francis Amasa
Walker / Library of Congress
122: Photographer unknown /
Library of Congress
128: T. W. Dibblee / United
States Geological Survey
131: Excerpts from T. W.
Dibblee / United States
Geological Survey
132: Google Earth (Overlay:
T. W. Dibblee / United States
Geological Survey)
140: Image courtesy Kirsten
Southwell
149: Russell Lee / Library
of Congress
162: Russell Lee / Library
of Congress
171: Russell Lee / Library
of Congress
182: Image courtesy
Colin Dutton
186: Image courtesy
Kirsten Southwell
196–97: Arthur Rothstein /
Library of Congress
199: James St. John / Flickr